REFLECTIONS
Service to our Country

2025

MILITARY WRITERS SOCIETY OF AMERICA

Senior Editor: Bob Doerr

Library of Congress Control Number: 2025910286

ISBN Paperback: 979-8-9988900-0-0

Published by the Military Writers Society of America
Weatherford, TX

Printed in the United States

Table of Contents

INTRODUCTION

The Military Writers Society of America (MWSA) is a diverse group of authors: active duty military, veterans, and civilians. We are historians, journalists, memoirists, bloggers, poets, musicians, and novelists. Some are old pros. Others are just beginning their writing careers. Most are somewhere in between. A love of the Unted States military binds us together as colleagues and friends.

Every couple of years, MWSA provides an opportunity for its members to be published in one of our anthologies. For some, it is the very first time their writing has become public. For others it is one more opportunity to tell their story. This anthology is no different. Thirty-one members submitted poems, articles, and stories. They are all included in this book. Some will make you laugh. Others you will find enlightening. You may even shed a tear.

Military service involves sacrifice and not only for the service member. Parents, spouses, and children also bear the burden. MWSA believes that writing can serve as therapy for some. Writing can also preserve history—one story at a time. We encourage our members to write, and we do what we can to improve our members' skills and ability to get published.

I hope you enjoy reading these stories. I know I did. The title and theme for this book is *Reflections: Service to our Country*. It's a broad topic and one that our members had a lot to write about. Whether it's a look back at the largest sea battle in history or a memory of a single flight in a fighter jet over Spain, the submissions included herein capture that theme.

I'd like to thank Jim Tritten, Terese Schlachter, Dwight Zimmerman, and Kathy Taylor for their assistance (volunteers all) in editing this anthology.

Bob Doerr, Editor

CHAPTER 1

THREE LIGHTS

BY RICHARD (ED) WOOTEN

1971

The U.S. Army Ranger School is eight weeks and three days of intense physical and mental challenges. Some days you're the leader, other days you're a follower, but at all times, you're trying not to fail. Failure is not an option.

Ranger School had three distinct phases—the Benning Phase which consisted of the City Phase and the Darby Phase; the Mountain Phase that was conducted in the mountains of Dahlonega, Georgia; and the Jungle Phase that was conducted at Eglin Air Force Base in Florida.

Each phase had to be successfully completed before moving on. We motivated ourselves by stating we earned two letters in the coveted RANGER tab for each phase. Thus far, we had completed the City Phase which earned the R. We were becoming accustomed, but not comfortable, to operating without sleep and food and constantly being on the move.

Completion of the Darby Phase would give us RA and the opportunity to move on to the Mountain Phase. We were near the end of the Darby Phase when we received a patrol order—a challenge to test our navigational skills. Our squad's mission was to escape and evade from point A and rendezvous with a local partisan at point B. We would find the local partisan among three lights at point B. The patrol order further stated that when we linked up with the partisan, we would tell him, "I have some mail for you." His response was to be, "Good, I hope it's a postcard." Upon successful execution of the challenge and password, the partisan would provide information concerning our extraction from the area.

There was a time deadline, and of course, it would take place under the cover of darkness. Standard operating procedures were to be followed, including safety protocols which included no contact with civilians, no navigating through the swamps, or traveling on paved roads or the railroad tracks.

Our squad disappeared into the darkness heading to point B. For this mission, I was a follower, another Ranger was the leader. We followed the compass' azimuth until we encountered a geometry lesson. Our line of progress intersected a road intersection just beyond a field with knee-high grass. Midway across the field, we spotted headlights of two vehicles approaching the intersection.

We hunkered down and waited for the vehicles to pass...but they didn't. They stopped and were joined within a couple of minutes by a third vehicle. Ugh! The intersection became a meeting point for three Ranger tactical officers in 1/4-ton vehicles.

Dammit! Two choices, stay hunkered down and wait for them to leave or develop an alternate route. We chose to wait'em out. This proved to be a very time-consuming decision.

When the tactical officers dispersed, we had expended almost an hour.

With our red-filter flashlights, we huddled around the squad leader and conducted a map check for a faster route. One of the faster routes passed through the swamp and followed the railroad tracks to close proximity of our check point.

I'm neither confirming, nor denying, that we chose a route that violated safety protocol; however, we eventually reached a landmark that indicated we were within a thousand meters of our destination. We had thirty minutes remaining.

We navigated the wooded, final leg of our journey alongside the river and spotted a man and a woman in a clearing near a pickup truck. We halted and observed. They were preparing fishing equipment, rods, and reels. There was a small campfire, a Coleman lantern, and a flashlight. *Aha, three lights—our partisan.*

Now, you may ask, "A woman was there?"

Yes, but this caused us no alarm. We were accustomed to training events that included encountering women and children. Early in our training, we learned these were Ranger cadre dependents and were role playing to add realism to our training.

We prepared to make contact with the partisan. In accordance with our standard operating procedures, the squad leader, accompanied by another squad member, would initiate contact. The rest of us would stay concealed and be prepared to "kill" the partisan in case things went badly.

The woman had walked toward the bank of the river and cast her fishing line. The man was still near the back of the pickup truck. With stealth, four of us moved closer to the man and took protective positions while maintaining our concealment.

The squad leader and his assistant emerged from the shadows, quickly took a half dozen steps alongside the pickup truck and approached the partisan's back. As the squad leader began, "I have some mail..." the man whirled and screamed, "What da hell?" and immediately slashed at the squad leader with his fishing rod.

Oh, shit, this is bad! We immediately opened fire with our M-16s (assault rifles). The man screamed louder and immediately started flaying his arms as if to cover the wounds that would occur from the 5.56 rounds. Of course, they were blanks, but he didn't know.

His wife also didn't know. She joined in his hysterical screaming and immediately ran into the water. I could have sworn she was walking on the water as she did her own version of escape and evasion heading across the Chattahoochee River.

We hauled ass, leaving Mr. and Mrs. Fisherman to their own accords.

We ran, single file, along the path. Before slowing to regroup, we made a slight left turn along the path. As we rounded the small bend, we came upon our tactical officers, standing around, smoking and joking, and surrounded by three small bonfires.

We told them about our other encounter. They laughed, said they shouldn't have been on post fishing, but said they'd have a helluva fish story for their friends. We were directed to the back of a deuce and a half truck, our transport back to base camp in preparation for deploying to the Mountain Phase.

I've often wondered if she made it across the river to Alabama, or maybe, she continued to Mississippi before she slowed down. Huah!

Ed Wooten received his commission in 1971 from Clemson University. During his career, he was a company commander with the 101st Airborne Division (Air Assault) at Fort Campbell, KY and was a battalion commander with the 2nd Armored Division at Fort Hood, TX. His awards and decorations include Ranger tab, Senior Parachutist Badge, Department of Defense Meritorious Service Medal, and Army Meritorious Service Medal (x4).

CHAPTER 2

HOMEWARD BOUND

BY MARK FLEISHER

I did not sleep much that night. That night would be my last night in Vietnam unless some unexpected and unwanted event occurred, keeping me from boarding the Freedom Bird that would carry me to the United States otherwise known as the World.

What could delay my departure after a year as an Air Force journalist? The base might come under attack. Enemy mortars might crater runways, preventing aircraft from taking off. I tried to clear my mind of such catastrophes.

A routine medical exam two weeks prior threw a scare into me. My white blood count was a bit out of whack, possibly revealing some sort of infection. During my year I religiously took my daily quinine and salt pills, and, except for a touch of what the docs thought was dengue fever, I avoided sick call. Thankfully, I was good to go after a second test produced normal numbers.

My flight left Tan Son Nhut at ten hundred hours – 10 a.m. in soon-to-be civilian time. I was ready four hours earlier. My duffel bag was packed. The rest of my belongings were crammed into a wooden crate the Air Force would deliver to my Brooklyn home.

I put on a new t-shirt, then my tan Class A's, freshly laundered and ironed by our barracks mama-san who appreciated the extra money she gladly accepted.

My buddy David scrounged a Jeep and drove me to the terminal which in those days looked like two Quonset huts joined together.

"Good luck, my friend," David said as he hoisted my duffel onto the pavement. "It's been real. Safe flight and happy landings."

"Take care of yourself, buddy. Keep your head low and your butt lower. I'll see you in January after you get back, and I want you in one piece."

David stepped back and gave me a snappy salute which I returned before executing an about face and marching into the terminal.

After a bored staff sergeant checked my paperwork, I boarded the World Airways 707 and scored an aisle seat just aft of the starboard wing. My seatmates were airmen second class, one from Alabama, the other from Illinois.

"Good to be going home, Sarge," the Alabamian drawled. "What y'all do over here?"

I told him I was an Air Force journalist, and he thought that was pretty cool. "It had its moments," I assured him. "How about you guys?"

Alabama drove a truck, mostly between Tan Son Nhut and Bien Hoa. Illinois was a sky cop – an air policeman. He nodded in the affirmative when asked if he'd seen combat during Tet.

An unseen voice wafted through the cabin, telling us the crew's preflight was done and we'd be leaving the gate in a minute or two. Quietly, slowly, the airplane pushed back, navigated the labyrinth of concrete aisles, and wheeled into takeoff position.

Engines at max thrust, rpms climbing, brakes released, we hurtled ahead. I silently counted...five seconds...ten seconds...fifteen seconds, now angling toward the blue vastness this late summer day.

That unseen voice returned.

"Ladies and gentlemen, may I have your attention. We have left the airspace of South Vietnam."

A tsunami of whoops and hollers washed over us all. Hell, it was like winning the lottery, the Super Bowl, and the World Series all at once.

"Y'all know," Alabama said, "my momma made sure I went to church every Sunday back home. She'd be prayin' all the time when I was in 'Nam. But extra prayin' when I told her I was leavin' on Friday the Thirteenth."

Illinois, who had been quiet, piped up. "I knew it was Friday the Thirteenth. Didn't want to say anything that might jinx us. What about you, Sarge?"

"Yeah, I felt the same way. Look, let's drop the Sarge. It's Nick." Illinois turned out to be Cliff, and Jesse was Alabama's handle.

We were about five hours from Guam's Andersen Air Force Base, home to the B-52s whose crews flew God-knows how many missions over Vietnam and left huge craters across the landscape.

I managed to sleep a bit, and when wheels hit the Andersen tarmac, a Texas twang shouted "Hot Damn! We're back in the good old U.S. of A."

"No, you dumb ass, Guam's not a state, responded a voice with a decidedly Latin tinge. "Estupido!"

"Who you callin' a dumb ass?" bellowed the Texan.

Great, I thought. Just what we need—a fist fight. Then the voice of reason, Jesse the Peacemaker from Alabama.

"Y'all knock it off now. Quit your arguin'. My friend Nick here is a writer so I'm guessin' he's pretty smart and can settle this darn thing. What do you say, Professor?"

I thought to myself...Think of Solomon and come up with something that at least sounded wise.

"Guam is not a state. But it is United States territory because, well, it is a Territory with a capital T. It's been a territory since we took it from Spain in the Spanish-American War. Everybody got that?"

7

A murmur of yeahs and yesses seemed to end the debate. I reminded everyone that our next stop was Hawaii and...

"I sure as hell know that's a state," interrupted Tex to a round of applause.

About a dozen passengers debarked in Hawaii, and we flew on to Travis Air Force Base located not far from Sacramento. Cliff, the air policeman, wished us all luck and walked into the crowd in the terminal. Jesse was more vocal.

"Nick, if y'all ever get to Cullman, Alabama, you come visit. Just tell anybody you're a buddy of Andy Fowler's boy. Everybody knows us."

Even if I doubted Jesse would ever come to New York, I wrote my name and phone number on a slip of paper. Hey, you never know.

I needed to be paid for nine days of unused leave time. But as it was around six in the evening, and the finance office was closed. Not a problem as my flight from San Francisco to New York didn't leave until two the next afternoon. I spent a night in transient housing, visited the finance office in the morning, and realized I had just missed a bus headed to SFO.

The next bus wasn't due to leave for another hour. I was feeling antsy and itching to get home when I spotted a taxi down the block.

"How much to SFO?" I asked the driver, a Black man in his fifties wearing a 49'ers cap.

I was stunned by his response.

"You kidding me? Only thirty bucks. Gotta be more than that."

If you are familiar with the Bay Area, you know SFO is well south of the city. From Travis the distance is more than 60 miles, probably a good 90 minutes considering the usual traffic.

Vietnam, I replied, when the driver asked where I was coming from.

"Damn, this ride would be on the house if I could do that. Get in, flyboy. We gotta get you home."

8

We arrived at SFO more than two hours before my flight. My driver and I got out of the cab, and as we shook hands, I put fifty bucks into his shirt pocket.

"Good luck to you, son. And God bless you."

"Thank you, my friend. Thank you."

I wanted to get something to eat, and a drink too. First, though, I wanted to call home, let them know my ETA in New York. The corridor leading to the restaurant contained a bank of telephones, all in use. I waited until one came available.

After a few minutes, a sailor finished his call, and a slew of nickels poured into his cupped hands.

"What gives?"

"This phone is screwed up," he said. "It's spitting out nickels when you hang up. It's all yours. Be my guest."

I called home, told Mom and Dad I'd be arriving around eleven that night. I said I'd take a taxi home. They had no part of that. And yes, the phone delivered a half dozen nickels into my waiting hands.

After a decent hamburger washed down with a bourbon and water, I found my departure gate and waited for my flight to arrive from Hawaii. Not long after, the airplane pulled up and began disgorging passengers, many bedecked with leis and Hawaiian shirts. Not a few of them looked like they had had one too many.

More leis and flowery shirts when I boarded. I found my seat toward the back of the plane and hoped to sleep for most of the nearly six-hour flight. I woke up somewhere over Iowa. I figured about another three hours or so.

That time passed uneventfully, and we touched down at JFK a few minutes after 11:00 p.m. I hustled off the airplane into the gate area and was hit by a couple of banners and a bunch of balloons there to greet other returning service men and women who had shared the flight with me.

I knew my mom and dad were not part of the hoopla. Not their style, nor mine for that matter. They spotted me, and mom rushed to greet me with a big hug and a kiss. Dad and I shook hands.

"You both look terrific," I said. "I'm good, even better now."

The taxi ride got us home around midnight. Running on adrenaline, I didn't feel sleepy. I wanted to stay up for a couple of hours and let it all sink in.

"You must be hungry," mom proclaimed. "What can I get you?"

"How about my favorite. You still remember?"

In a matter of minutes, tuna with just enough mayonnaise flecked with diced apple and red onion appeared between two slices of rye toast.

I stared at the sandwich for what seemed an eternity.

Mom looked at me, her concern evident.

I snapped out of my trance.

"I'm home, mom. Really home."

Award-winning Albuquerque writer Mark Fleisher has published six books of poetry and prose. His recent book—*Persons of Interest*—contains a Baker's Dozen of poems and stories. His fifth book—*Knowing When*—was awarded a 2024 bronze medal for poetry by MWSA. Fleisher, an Ohio University journalism graduate, served in the United States Air Force. He spent a year in Vietnam as an Air Force combat news reporter and received a Bronze Star for Meritorious Service.

CHAPTER 3

CITIZENS MILITARY TRAINING CAMP

BY RUDY COLOMO VILLARREAL

Nineteen-thirty was the first year that my father Rodolfo (Rudy) attended the Citizens Military Training Camp (CMTC) at Camp Stephen D. Little near Nogales, Arizona. These camps were voluntary and were established to provide military training for high school age boys during the summer months. Mr. Harry Gerdes was the vocational teacher at Morenci High School in Morenci, Arizona, where he promoted the CMTC to many of his students. He was also a reserve Army officer and would later play an important role during WWII when he recruited many of his former students, including my father, for civil service jobs in San Diego, California.

Vocational education was the predominate curriculum that Rudy and his schoolmates were exposed to. They were taught practical shop skills such as working with metal or wood. This increased their chances of obtaining the better-paying jobs. Very few from the mining towns went to college in the 1930s, and it would be several decades before that changed.

During their first summer, everyone trained with the infantry, and thereafter you had a choice of infantry or cavalry training.

Although still in existence in the 1930s, the horse cavalry was in its final years and would be phased out during the 1940s after WWII. Rudy's first summer of training was with the 25th Infantry Division at Camp Little. There he met boys from other towns and cities throughout Arizona. One of those was Robert Manheimer, an acquaintance from Metcalf, Arizona, who was then living in Superior, Arizona. His sister Virginia was a good friend of the Villarreal family. Robert would go on to serve in WWII as an infantry officer and make a career of the military, rising to the rank of Lieutenant Colonel. One of the young staff officers that Rudy remembers was a second "louie" by the name of Barry Goldwater.

During his second summer, Rudy chose the cavalry and returned to Camp Little to train with the 10th Cavalry. This unit, made up of African American troops, was famous for serving during the Indian wars of the nineteenth century and the hunt for Pancho Villa during the punitive expedition into Mexico in 1916. His final summer was with another famous unit, the 7th Cavalry, at Fort Bliss, Texas. This unit, while serving during the Indian wars, is best remembered for Colonel Armstrong Custer and his loss in the Battle of the Little Bighorn in 1876. One of Rudy's memorable experiences that summer was saving ticket fare by riding the rails on a freight train from El Paso, Texas to Lordsburg, New Mexico where connections were made to Clifton near his hometown of Morenci. Others from Morenci who trained with the cavalry were John Ainza, Joe Catero, and Henry Velasquez.

This is a photo of my father Rodolfo, while training with the CMTC at Camp Little around 1931. He is dressed in a cavalry uniform. How I acquired this photo is an example of the many "small world" encounters I have experienced in my life. In the early 1980s, my wife Mary Ellen and I were on vacation visiting her mother Jessie who lived in

Salinas, California. We were invited to dinner by Jessie's cousin, Concha Montoya, who lived in the nearby town of Watsonville. I had never met the Montoyas, and after dinner while we sat there chatting, Concha's husband, David, pondering over the Villarreal name, asked me if my father had ever been in the cavalry. I told him that I thought he had trained with some cavalry units when he was in high school. David excused himself from the table and returned in a few minutes holding a photo album. He pointed to one of the photos and asked, "Is this your father?" I looked at the photo in amazement and answered, "No, that's my uncle Froilan, but this other photo is my dad." It was an incredible moment and was the highlight of the evening. David had taken the photos fifty years earlier and had not recorded the names. He remembered that the boys were from Morenci, a copper mining town similar to his hometown of Hayden. When I presented copies of the photo to my father and uncle, were they ever surprised! And they both did remember David.

Photo of author's father
(Courtesy of David Montoya)

Rudy Colomo Villarreal is the author of *Arizona's Hispanic Flyboys 1941-1945*, published in 2002 with a new edition in 2016. More recently, he is the author of *Rare Bird: Hispanic Military Pilots of the USA*, published in 2020. On August 27, 2022, *Rare Bird* was awarded a silver medal by the Military Writers Society of America at their annual awards banquet held in New Orleans, Louisiana.

A graduate of Northrop Institute of Technology, he is retired after thirty years in the aerospace industry, having worked for Douglas, Lockheed and Garrett Airesearch. He served in the U.S. Army from 1964-1966. Rudy and his wife Mary Ellen live in Tempe, Arizona.

Photo by Rudy's granddaughter, Libbi Magat.

CHAPTER 4

REFLECTIONS ON
WORLD WAR II HEROINES

BY PENNY RAFFERTY
HAMILTON, PH.D.

During World War II, patriotic and talented American women contributed to winning the war. Many serving first in the Women's Army Auxiliary Corp (WAAC) which morphed into the Women's Army Corp (WAC), Navy WAVES, Coast Guard SPARS, and Marine Corps non-combat roles added much needed "woman power" to vacant support slots at home.

Lesser-known units as the Central Postal Directory Battalion 6888th (nicknamed The Six Triple 8) served in Birmingham, England. Recent books and films document the story of this female, mostly Black American unit. Led by the charismatic and pragmatic, 26-year-old Charity Adams Earley, 850 women with the mission to sort and deliver years of backlogged mail intended for our Troops in the European theater, processed an average of 65,000 pieces of mail each shift. The women of the 6888th were given six months to sort the millions of pieces of mail stored in airplane hangars. They did it in only three months. Finally, in 2022, these trailblazing women were awarded the well-deserved Congressional Gold Medal.

These women serving in the Central Postal Directory Battalion 6888th were part of the only Black American female unit to serve overseas during World War II. (Library of Congress)

U.S. Army Major, Charity Adams Earley, was the first Black American woman to be an officer in the Women's Auxiliary Army Corps in WWII. Adams Earley was the highest ranking African American woman in the Army at the end of WWII. (National Archives)

U.S. Army Air Force Winged Angels

World War II crew member flight nurses provided life-saving care in thirty-one medical air evacuation squadrons in Europe and the Pacific. On February 18, 1943, the first class of thirty-nine graduated from the U.S. Army Air Forces Flight Nursing Program at Bowman Field in Louisville, Kentucky. By the end of the war, around 500 "Winged Angels" had provided life-saving care. They cared for over one million military patients evacuated by air between Jan 1943 to May 1945 with few lives lost during the medical transport.

"The role of the flight nurse during World War II was revolutionary. The lives of the wounded rested on their shoulders-from starting IVs and oxygen to keeping all the ill and injured calm through a combat medical evacuation," according to the *DAV Magazine* July/August 2024. Women in Aviation International (WAI) honored our U.S. Army Air Forces Flight Nurses with induction into their 2024 Pioneer Hall of Fame.

Women's Auxiliary Ferrying Squadron (WAFS)

On September 10, 1942, with the support of the U.S. Air Transport Command, experienced aviator, Nancy Harkness Love, was appointed Commander of the WAFS. Their purpose was delivery of new airplanes from the factory to U.S. military bases. The female pilots were Civilian Civil Service Employees, who paid for their WAFS uniforms. They paid their own transportation costs to the WAFS Delaware Headquarters at New Castle Army Air Base. The WAFS were recruited from among commercially licensed American women pilots with at least 500 hours flying time and a 200hp rating. WAFS pilots who were selected actually averaged about 1,100 hours of flying experience. Many were also flight instructors.

Nancy Harkness Love, WAFS Commander, ferried new airplanes, including the B-17 Flying Fortress, in support of the American military in World War II. (Smithsonian)

During the WAFS existence, about twenty-eight women flew. Many continued to make aviation history: Nancy E. Batson, Bernice Batton, Delphine Bohn, Cornelia Fort, Barbara Erickson, Teresa James, Gertrude Meserve, and Evelyn Sharp. Although their original mission was to ferry U.S. Army Air Corps trainers and light aircraft from the factories, soon they were delivering fighters, bombers, and transports. Betty Gillies and Nancy Love became the first women to ferry the Boeing B-17 Flying Fortress, four-engine heavy bomber. The WAFS members made significant contributions toward winning World War II.

The first woman to qualify as a WAFS pilot was Betty Gillies. (U.S. Air Force photograph)

Women Airforce Service Pilots (WASP)

In August 1943, a new organization was created: the Women Airforce Service Pilots (WASP). The WAFS were merged with the Women's Flying Training Detachment (WFTD), which was under the leadership of Jacqueline Cochran. Many of the original WAFS members joined the WASP. Although flying under military command, the WASP were classified as civilians and paid through the civil service. The WASP paid their own transportation to Avenger Field in Texas for training. They were charged room and board which was deducted from their monthly pay.

Women Airforce Service Pilots ferried 12,000 new and repaired aircraft to U.S. military bases and repair stations. They flew military personnel from 120 Army Air Bases across America. They towed targets for men on the ground who were honing their gunnery and anti-aircraft skills. WASP supplied the womanpower to free up manpower to fight in combat. WASP flew over 60 MILLION miles in military planes. WASP flew almost every kind of aircraft needed by our Armed Forces during World War II.

WASP Dorothy Bancroft (Hammett) flew the PT-17, BT-13, AT-6, and the UC-78. Invited by Jacqueline Cochran to join the WASP, Dorothy Bancroft originally flew new planes from the manufacturers to U.S. airfields. According to her granddaughter, Kelly Mackura, Bancroft later was asked to instruct military men for instrument flight. If male pilots were offended, the WASP repeated the mantra used in World War I when The Flying Schoolmarm, Marjorie Stinson, taught Canadian men to fly—"A woman taught you to walk. A woman can teach you to fly."

When Jacqueline Cochran called for patriotic women to apply for the WASP, over 25,000 American women answered. Only 1,879 women were selected for the 27-week training at Avenger Airfield. During the 16-months of the WASP existence, eighteen classes graduated, yielding 1,074 highly trained pilots. WASP Barbara Erickson London flew a five-day, 8,000 mile air mission. First, she delivered a DC-3 from her Long Beach, California base to Fort Wayne, Indiana. There, she picked up a P-47 Thunderbolt for delivery to California. Next, Barbara flew a P-51 Mustang back to Fort Wayne where she picked up yet another

Women Airforce Service Pilots, Frances Green, Margaret "Peg" Kirchner, Ann Waldner, and Blanche Osborn, leave their B-17 Flying Fortress aircraft, "Pistol Packin' Mama," during their ferry training at Lockbourne Army Airfield, Ohio, 1944. (U.S. Air Force archival photograph)

P-47 to fly to Long Beach again. On the last leg of this whirlwind mission, she flew a P-61 Black Widow to Sacramento.

On April 1, 1943, the Grand Island Nebraska airfield (GRI) was activated to train bomber crews. According to the Texas Woman's University archive, three WASPs were assigned there. One of these women was Eleanor Fairchild Stebbins. Her nickname was "Fearless Fairchild." At only 19, she was the youngest pilot in her class, 44-W-6. Classmates, Evelyn McNulty Perrin and Lorraine Lasswell, joined her on the newly expanded airfield.

An interesting WASP and aviation history story happened at GRI. The new B-29 Super Fortress, built in Omaha at the Glenn L. Martin factory, rushed to production, was the newest, biggest, and most complicated bomber. Many pilots felt it was unsafe, especially with frequent engine fires. Built with Wright engines, pilots quipped the B-29 had the "wrong" engines. In the summer of 1944, Lieutenant Colonel Paul W. Tibbetts (yes, the famous World War II pilot who later flew the B-29 Enola Gay, named for his mother, to Japan with the atomic bomb) personally invited two outstanding WASPs, Dora Dougherty Strother McKeown, 43-W-3, and Dorothea "Didi" Johnson Moorman, 43-W-4, to train with him in a B-29. Long story short, after three days of tutoring the WASPs, Tibbetts wrote a letter of invitation to the male pilots to see the new B-29 pilots fly Ladybird. Both WASPs flew and landed the behemoth before large gatherings of reluctant male pilots. They patiently answered many operational questions. Dora and Didi flew crews on orientation flights. And, as they often say, "the rest is history…or maybe herstory!"

The "her-story" of WASP service includes thirty-eight killed in training or on air missions. Because WASPs were considered civilians, the military did not pay for their remains to be shipped home to their families for burial. Often, the other WASPs donated money if the family could not afford the cost. Those thirty-eight families were not even allowed to display the traditional Gold Star in their window, indicating a family member lost their life while serving our nation in the armed forces. In 1977, it literally took Congressional action for WASP to retroactively have veteran status and benefits. It was not until 1984, almost 40 years after the end of World War II, that WASPs were awarded their World War II Victory Medal. For those serving more than a year, they were awarded the American Theater Medal. In 2010, WASPs were awarded the Congressional Gold Medal, the highest civilian expression of national appreciation, for their World War II contributions. Today, the stories of our World War II WASP continue to inspire.

Thirty years ago, then Secretary of the U.S. Air Force, Sheila Widnall said, "I am proud to recognize the contribution the WASP made to World War II. They set the stage for today's women to fly and fight with their spirit and enthusiasm. These heroines…heard the call and responded with the skills and dedication that gave our country the boost it needed to win World War II." Reflecting on the many heroines of World War II is an inspiring exercise for all.

Multi-award winning aviation and women's history writer, Dr. Penny Hamilton is a Laureate of the Colorado Aviation, Women's and Authors Halls of Fame. She co-holds a World's Aviation Speed Record with her husband, William Hamilton. Named to the Amelia Earhart Forest of Friendship, she is winner of the Columbia College Distinguished Alumna Award, University of Nebraska Alumni Achievement Award, and two-time winner of the Regional U.S. Small Business Media Advocate Award, learn more at www.PennyHamilton.com

CHAPTER 5

HUMBLED:
A TRAVEL JOURNAL

BY TRUEST CARTER

Normandy, August 1998

I'd drawn bright yellow highlights mapping our route to the American Cemetery Normandy. Hailing from rural Appalachia, many obscure European destinations bunched together like one big multi-country. I'd left my little town for active duty service, and while overseas, I discovered the plethora of countries and cultures that formed the European continent. So, when Grandma Dot and Cousin Erma learned I would be traveling to France, they volun-told me to look for one particular soldier. I had a reunion to arrange.

Months before, when the nurse manager suggested that I spend my last two years on active duty across the pond, I weighed my options: go home to the familiar hills of Pennsylvania or explore Europe. Not surprisingly, within two months, I was stationed at RAF Lakenheath (Royal Air Force Base Medical Center near Cambridge, Great Britain). Almost upon landing, I'd made friends with a carload of nurse and medical corps officers eager to take in historic sights and extraordinary landscapes.

One adventure, spurred on by family history stories from Grandma Dot and her cousin, Erma, resulted in the discovery of an eighteenth century Staffordshire marriage certificate in the basement of an old...no...ancient church. In my letters home to Dot and Erma, I'd described the scene from the microfilm room when I jumped up and yelled, "I found them," and other researchers joined in my celebration. Unsurprisingly, the Staffordshire coal miners were ancestors of the dust-covered Pennsylvania miners of my lifetime. I guessed they emigrated to find a new life only to take the same employment. Except in the sprawling United States, they could buy their own homes and pass the land down for generations. As the years passed by, the entrenched Pennsylvanians rarely traveled and never to Europe.

I shouldn't say never. A generation from the 1940s earned a free trip that cost many their lives. Erma recalled how, soon after the birth of her second child, her husband had gone to war. He had not returned. Private Helmes had been buried sometime after the invading Allies pushed back German forces. She shared scant details. Erma never remarried.

On my Normandy excursion, an enthusiastic carload of my new Air Force friends joined in the adventure. Sherry turned up the FM volume, and we sang along to Boyzone's new hit *No Matter What*. Well, I mouthed the chorus—*No, no matter*. No one wanted my tone-deaf vocals. When the crowded car made a sharp turn onto the Rue des Allies, the three of us in the back seat shifted shoulder to shoulder.

Monet could have painted the morning haze as it coated every church spire and golden field in Normandy. The August day sweltered, but the ocean breeze kept us comfortable. English signs read "Utah and Omaha Beaches" with straight arrows. Could have guessed those were the American landing spots. *Wonder what they were named before the invasion.* Gradually, the idyllic scenery became littered with strange concrete bunkers, ship-sized debris, and waves crashing onto abandoned massive metalworks.

"There," I pointed at the car park. "Let's snap some pictures."

"Did everyone have a grandfather who served?" Diane asked while she stuffed her backpack.

My friends shared stories of grandparents who braved the Pacific front while others had waivers. My friends had gone with me on other family history trips, mostly to explore new towns, but it hit me rather abruptly that they had no connection to Normandy. Maybe I'd taken it for granted because I'd heard about my grandfathers' and great-uncle's World War II antics on the Atlantic and Pacific.

Sherry pulled her hoodie tighter. "Can you imagine being dumped on a beach under fire?" She was always cold, even in August, but perhaps another feeling caused these chills.

We made our way, wide-eyed and in awe, to the beach walkway, and I realized how far out of their element those miners-turned-soldiers had been. There were no rocky outcroppings or trees to provide cover. The vast, empty beach humbled and quieted the noisy Americans.

Definitely could not imagine. Our military training consisted of quickly unpacking pallets of portable medical supplies. Highly technical and gas-impermeable trailers were stocked to support our efficient trauma unit. The mock battles and terrorist scenarios (in the picturesque English countryside) couldn't compare to what the medics of World War II must have faced on those beaches more than fifty years prior. Through the decades, sounds of gunfire and mines echoed off the ocean waves.

Diane leaned over the railing. "God, all those souls. Such a sad place. The air even feels heavier here."

It did. It really did. The sandy coast, still littered with anti-landing "Czech hedgehog" barriers from the largest amphibious assault in history (in all of humankind), had no resemblance to a typical beach anywhere else in my known world. No families spread out blankets on the hallowed ground. No rainbow of umbrellas was posted in the sand. Elaborate historical markers accounted for the thousands lost in each landing zone. We breathed it in and said little.

In no mood for lunch at a French cafe, we left to finish our film rolls at the cemetery. I cracked the window and breathed in the sea air on the quiet drive from the shore, through the first town to be liberated, and up to grass greener than inside Pittsburgh's Three Rivers Stadium.

Diane glanced back from the front seat. "You think you'll find your cousin?"

I'd forced the car-full to learn about the sad case of Private Helmes. They half listened. Travel guide tidbits about Mont Saint-Michel held their interest instead. For some, Normandy provided a travel "been there" checkmark. For me, it was an immersion in the spirit of it. Back home, I couldn't relate to my generation of cousins who had never known their grandfather. He was like a legendary character from a tragic war story. Yet, I was the lucky one who got to see the beaches that his infantry had stormed. That young man, who had sacrificed ever being called "Grandpa," rested nearby.

And there I was, fifty years later, with a group of American officers stopping for baguettes without any conflict...unless you count the eye-rolling French ladies and flirtatious French men. Green and golden fields passed between white travertine Gothic churches rising from humble villages. American Cemetery signs interrupted the landscape.

"I hope his marker is in a nice spot. Grandma and her cousin wrote stories about family history. They don't say much about her husband, though. I only found out he was buried in Europe after I PCS'd here." Maybe her fifty years of widowhood had blurred reports on long-dead facts.

Sherry sorted through her souvenir bag. "She probably didn't want to talk about it. He must've been very young."

"Yes, younger than I am now." I leaned onto the headrest. *Can't imagine what they suffered.*

Diane brushed her hair into a ponytail. "Wonder if you'll really find him."

I creased my brow. "Hope so." *Did she not think I had a cousin buried here?* Maybe I'd gone on about him too much. I couldn't wait to snap a solitary photograph of his etched cross and send it back home to Grandma Dot and Cousin Erma.

The sun burned off the remaining haze over the fields of grave markers. Rows of white marble crosses and stars stretched for acres, or football fields, as Americans like to compare for size. *Endless football fields.* I sneezed at the newly mowed patch around the flagpoles. Crowds at the visitors' center bumped elbows and maneuvered for space in lines. Travelers took time out of their culinary, artistic, or otherwise historical French sightseeing to pay respects. The lines moved quickly, and I approached the clerks. Diane stayed with me while our friends wandered around the landscaped paths.

The staff offered grave searches and provided directions on the grids. It was too immense an area to search for individual names on stone. Thank goodness for the nearby tower desktop computers.

I pulled out Grandma Dot's letter with Private Helmes' details. Her cursive handwriting on ruled paper had become more squiggled as she aged. It was legible, and seeing it made me wish she could have been there. We searched for her cousin, after all. She had youthful memories of the boy who married Erma and then left forever.

"Here, hurry." Diane signaled me over to the next clerk. She kept close by my side. Either from curiosity or doubt, but it felt more like doubt. "Tell her the name."

I stood tall in front of a young American woman with braided blonde hair when she read my grandmother's writing and typed in the name. A minute or so of circles on a thinking computer screen resulted in one match.

"Yes, I found one infantry soldier with that name—from Pennsylvania." She pulled out a larger map, folded it at least eight times, and smoothed a section from corner to corner.

Diane leaned over me toward the screen, either proving to herself or checking my facts, I didn't know. Unless…maybe she was a little jealous. Jealous that I had a connection to this amazing history. With over 400,000 American war dead, I could claim one. It wasn't just about pride in American historical connections; it was about relating to soldiers our own age who gave all. I happened to be related to one. It only took the one, though, to mourn and empathize with a widow and her children from fifty years before. One story. Humbled.

"She found him. Cool." I gripped the letter. I couldn't wait to write all about it. *Hope it's not far to walk in this sun.* Although, I would have walked all the football fields for that picture.

"Only…," the lady pursed her lips. "He's in an American Cemetery in Belgium. Here, I'll write the location on this map."

"Belgium? My family thought for sure it was after D-Day." I stood my ground.

"No, ma'am. He is in Belgium."

Diane lowered her head.

I gulped in a breath. "I see. Thank you." A good bit disheveled inside, I wasn't ready to move on. "May I make a donation?" I unzipped my fanny pack.

"No. We aren't allowed to accept donations." Her look reminded me that the price had already been paid on those beaches.

How could my family have it so wrong? Even his widow told us France. If he'd been injured, the dates made sense for his passing after the invasion.

Diane and I pushed through the crowded center and out to the manicured path.

"So embarrassing." I looked over the letter again.

"Nope." Diane gripped my shoulder. "It means he survived this. He made it to Belgium. Now, we have another destination to plan."

I exhaled with a whoosh and nodded. "Bet Belgium is beautiful in the fall."

Truest Carter, a 20-year resident of Northern Virginia, is a USAF veteran, nurse, and writer of many genres. She has a BSN, an MA in Liberal Studies, and an Emory University Creative Writing Certificate. An award-winning poet, she has also published fiction in journals and online serials. When she is not researching a story or attending writer events, Ms. Carter recharges on hikes with her husband and their energetic rescue dog.

Traci Carter writing as Truest Carter

CHAPTER 6

DRAMA AT LEYTE GULF

BY JIM TRITTEN

O n October 23-26, 1944, the largest naval battle ever fought took place east of the Philippine Islands—the Battle of Leyte Gulf. The American Third and Seventh Fleets, with an Australian Task Force, engaged the Imperial Japanese Navy in an epic struggle to the death. On October 20th, the U.S. landed 130,000 troops of the U.S. Sixth Army on Leyte Island to fulfill General Douglas MacArthur's promise to return to the Japanese-occupied islands taken by force in 1942. The Japanese mounted an immediate counterattack by 70,000 soldiers of their occupation ground forces against the U.S. Sixth Army. Their ground operation was coordinated with an all-out effort by the Japanese fleet to destroy supporting Allied naval forces offshore in Leyte Gulf. The Allies' control of the seas was critical to the invasion's success.[i]

While General MacArthur's landing on Leyte on October 20th, accompanied by the president of the Philippines, was a significant event, the true heroes of this battle were the ordinary sailors and airmen of the U.S. and Australian navies. These brave men, in a display of unparalleled courage, laid down their lives in a valiant effort to repel the Japanese counterattack. As one commanding officer eloquently said, they fought "against overwhelming odds from which survival could not be expected."[ii]

Multiple confrontations occurred over thousands of square miles of the Philippine Sea during the Battle of Leyte Gulf. One of these, the meeting engagement off Samar Island, involved Task Unit 77.4.3 (radio call sign Taffy 3).[iii] It was one of four significant actions fought by Allied seamen and flyers. Samar, the center of gravity of the entire Leyte action, held the fate of the whole operation. If the landing failed, the entire operation failed, and it would not matter how many ships had been damaged or sunk. Their actions would become known as the Battle off Samar Island and forever be recognized as one of the most memorable days in the combat history of the U.S. Navy.

Taffy 3, a minor task unit, was equipped with small, unarmored escort carriers supplemented by old, lightly armored destroyers and thin-skinned destroyer escorts. They were not equipped with the most current technology, and their initial expectation was to engage in airstrikes and naval gunfire against shore targets—not to face battle-hardened enemy warriors in ships that had previously wreaked havoc on the Allies. This small unit was only intended to support the landing force directly. However, the Americans onboard the ships of Taffy 3, which directly defended the landing area, were about to face an unexpected and formidable challenge.

On October 25, 1944, Imperial Japanese Navy Vice Admiral Takeo Kurita surprised the Americans off Samar Island to the north of the landing area, with the largest surface battle fleet fielded by Japan since the 1942 Battle of Midway. Kurita's fleet consisted of four battleships (including the giant *Yamato* with its 18.1-inch guns), six heavy cruisers, two light cruisers, and eleven destroyers. They were supported by shore-based aircraft from Luzon.

Only U.S. Rear Admiral Clifton A.F. "Ziggy" Sprague and Taffy 3 stood between this overwhelming force and the support ships in Leyte Gulf. His command consisted of six small escort carriers, three destroyers, and four destroyer escorts. Hardly an even match. Escort carriers, also known as "jeep carriers," had a small flight deck and a limited complement of airplanes. Sprague ordered his flattops to launch their aircraft and have them retire to a rainstorm to the east. Sprague's surface ships were ordered to make smoke to hide the carriers. All the major combatants of the U.S. Navy were otherwise engaged to the north and south and could not join the battle.

The nearest destroyer to the Japanese forces was the USS *Johnston* (DD-557),[iv] commanded by Lieutenant Commander Ernest E. Evans, U.S. Navy. When Evans saw the pagoda masts of enemy battleships and cruisers on the horizon, he first laid a smoke screen as ordered and then, on his own initiative, steered his small and outclassed ship directly toward the enemy. David vs. Goliath. An action that would make Horatio Nelson proud: "Never mind manoeuvres, always go at them."[v]

After closing within firing range, the *Johnston* fired some two hundred 5-inch artillery shells and all ten torpedo tubes, striking the enemy cruiser *Kumano*. The *Johnston* then took evasive action and ducked into a rain squall. After receiving hits on the bridge and elsewhere, *Johnston* received a general order from Sprague for all destroyers to make torpedo attacks. *Johnston* rejoined the fray, making virtual torpedo attacks (the Japanese defenders did not know *Johnston* was out of torpedoes) and fighting furiously with her modest complement of 5-inch guns. *Johnston* and other destroyers attempted to draw enemy fire and force the Japanese ships to take evasive action—whatever was necessary to keep the Japanese forces away from the transports, supply ships, and the landing force.

The 2,100-ton *Johnston* next took on the 30,000-ton battleship *Kongo*, scoring fifteen hits before ducking into her own smoke screen in safety. During the enemy counterattack, Evans lost two fingers and, in the force of a blast, all the clothing above his waist was blown off. When an enemy cruiser engaged one of the escort carriers, Evans closed on the cruiser and scored four hits with his guns. As a squadron of four Japanese destroyers and a light cruiser maneuvered to box in the escort carrier, Commander Evans seized the initiative by attacking the whole squadron. *Johnston's* ferocious close-in gunfire so startled the enemy that their torpedoes were launched prematurely, temporarily sparing damage to the escort carriers.

Kurita singled out *Johnston* for his vengeance. The ship was hit by so many rounds of enemy gunfire that the crew could not plug the holes fast enough. After a series of mortal blows to the destroyer, Lieutenant Commander Evans ordered the crew to abandon the ship. *Johnston* slipped beneath the waves, taking 186 of its 327-man crew with it. Evans' body was never recovered. As *Johnston* slipped beneath the waves, the Japanese destroyer *Yukikaze* passed close by, and her captain saluted his worthy adversary.[vi]

The crews of many other ships that fought that day off Samar Island and throughout the Battle of Leyte Gulf also displayed extraordinary heroism. Aircraft initially ordered to safety joined the fray and made attack after attack against Japanese battleships and cruisers. They had been loaded with ammunition designed to support the landing—not to engage armored warships. Even after their ammunition was exhausted, pilots from the small escort carriers continued to make attack runs into the withering teeth of the antiaircraft fire of enemy surface forces— attacks intended to force a reaction by the enemy and thereby upset opposing fire and maneuver. The Japanese seamen did not know that the aircraft diving on them may have already expended their ordnance.

Most pilots and ship companies that fought at Samar that day were reservists with scant combat experience other than against submarines and shore batteries. Specifically, the crew of the *Johnston* consisted of mostly married draftees. Eighty-five percent of the *Johnston* crew was "green," and only one-third of her officers had ever been to sea.

After facing what he thought were stronger forces than his, Kurita withdrew his fleet. Only two of his five battleships remained combat-capable when the Japanese returned to their bases. Three Japanese heavy cruisers had been sunk, and the other three were damaged.

American losses were considerable, but the objective of protecting the landing force, troop, and supply ships had been achieved. Two escort carriers went to the bottom, as did two destroyers and one destroyer escort. The four remaining escort carriers were all damaged, as were the remaining one destroyer and two destroyer escorts.

In total, 1,161 American seamen lost their lives, and another 913 were wounded. These losses are comparable to U.S. casualties at the Battle of the Coral Sea and the Battle of Midway combined. Fleet Admiral Chester Nimitz, Commander in Chief of the U.S. Pacific Fleet, wrote afterward that the success of Taffy 3 was "nothing short of special dispensation from the Lord Almighty."[vii] Historians have cited The Battle off Samar as one of the greatest last stands in naval history[viii]

The doctrinal issue of how to best protect a landing force reared its ugly head in the Battle of Leyte Gulf. On the one hand, some wanted the Navy to remain close to the landing force to provide direct protection to vulnerable ground forces on the beach and en route on fragile landing craft. This was the role that Rear Admiral Sprague's small force, Taffy 3, played that fateful day. On the other hand, others wanted to leave the landing area and seek out the main battle fleet over the horizon. That was the role that Admiral William Halsey's Third Fleet played—seeking contact with major Japanese aircraft carriers far to the north and abandoning Taffy 3 and its landing forces to their fate. Unfortunately, the Japanese force to the north was a decoy, and Halsey fell for it.

This doctrinal issue has been played out numerous times in history. Most recently, in 1942, the naval escort fleet withdrew from the landing area at Guadalcanal. Industrious Marines who were left behind without the Navy created the *Faciat Georgius* [Let George Do It] medal with a depiction of an admiral dropping a "hot potato" into the hands of a Marine on one side and the "Shit hitting the fan" on the other.[ix]

Such behavior replayed itself off Samar Island in 1944. The quality of the disparate forces set the stage for heroics that will forever hallow the halls of U.S. Navy valor. On paper, Taffy 3 was no match for what they encountered that one October day. However, it was not the admirals who cunningly deployed their forces in a way that ensured MacArthur's landing in the Philippines succeeded. Instead, it was the courage and determination of ordinary seamen and aviators whose equipment was hopelessly outclassed who gave their last ounce of blood to ensure the landing force was not destroyed. Taffy 3 was awarded the Presidential Unit Citation, and a total of twenty-two Navy Crosses were presented to specific individuals from the unit for actions off Samar Island.

Lieutenant Commander Ernest E. Evans, Commanding Officer of the USS *Johnston*, was a full-blooded Cherokee Indian. He first served in the Oklahoma National Guard and then entered the U.S. Naval Academy, graduating in 1927. On September 28, 1945, Lieutenant Commander Evans was posthumously awarded the Medal of Honor[x] for his inspirational leadership under the most severe combat conditions. He was the first Native American in the U.S. Navy to earn the Medal of Honor.

What made Commander Evans such an exceptional combat leader that the nation would recognize his efforts in such a manner? Was his combat leadership influenced by the difficulties he must have incurred because of his minority status at school, in society, or the military? Was his desire to lead by example shaped by his experiences while in the National Guard and at the Naval Academy, or was it learned from some mentor in the fleet?

Did the exploits of John Paul Jones, British Admiral Horatio Nelson, or the heroes of his Native American culture inspire Commander Evans? Did his crew view Evans as a descendant of such famous admirals, or was he inspired by tales he heard as a boy of Cherokee war chief Dragging Canoe?[xi] Had he followed Horatio Nelson's practice of first talking with his wardroom about what they could expect in battle? Did any of Evans' officers question his decisions that fateful day? His gunnery officer, who survived the engagement, said of Evans' leadership "...that it made us all willing to follow him to hell."

Leadership was defined in the *War Instructions: United States Navy, 1944,*[xii] as: "...the art of inspiring, guiding, and directing bodies of men so that they ardently desire to do what the leader wishes." This is what the commanding officer of USS *Johnston* did when he charged at the enemy battle fleet with his outclassed, outgunned David against the behemoth pagoda-masted battleships of Admiral Kurita's fleet. We will never know what inspired Lieutenant Commander Evans to greatness.

Did it matter who commanded USS *Johnston* on that fateful October 25, 1944? Are inspirational leaders born out of a crisis? Can an external event trigger individuals who have not exhibited any particular "charismatic" traits? Are heroic traits latent in us all and merely await awakening? Are there ways the U.S. Navy can better prepare its future combat leaders in today's fractured nation without consensus on its role on the international stage?

How do we develop such distinguished combat leaders in the future? How do we find such men and women who can inspire sailors and naval aviators to greatness in the face of combat at sea when their country again calls? Can such combat leaders be cultivated, or are they born and hopefully recruited into naval service when needed?

The Medal of Honor citation for Lieutenant Commander Evans reads as follows:

"For conspicuous gallantry and intrepidity at the risk of his life above and beyond the call of duty as Commanding Officer of the USS *Johnston*, in action against major units of the enemy Japanese fleet during the Battle off Samar on October 25, 1944. The first to lay a smoke screen and to open fire as an enemy task force vastly superior in number, firepower and armor rapidly approached, Commander Evans gallantly diverted the powerful blasts of hostile guns from the lightly armed and armored carriers under his protection, launching the first torpedo attack when the *Johnston* came under straddling Japanese shellfire. Undaunted by damage sustained under the terrific volume of fire, he unhesitatingly joined others of his group to provide fire support during subsequent torpedo attacks against the Japanese and, outshooting and outmaneuvering the enemy as he consistently interposed his vessel between the hostile fleet units and our carriers despite the crippling loss of engine power and communications with steering, shifted to the fantail, shouted steering orders through an open hatch to men turning the rudder by hand and battled furiously until the *Johnston* burning and shuddering from a mortal blow, lay dead in the water after three hours of fierce combat. Seriously wounded early in the engagement, Commander Evans, by his indomitable courage and brilliant professional skills, aided materially in turning back the enemy during a critical phase of the action. His valiant fighting spirit throughout this historic battle will endure as an inspiration to all who served with him."[xiii]

In October 2019, a wreck was found at the bottom of the Philippine Trench at a depth of 21,180 feet. It was not until March 2021 that the wreck was positively identified as the *Johnston*. She had been broken in half after absorbing a 3,000-pound artillery shell from *Yamato*. Johnston's hull number, 557, is still visible, white against gray, buried in the sand with much of her crew entombed forever at their battle stations.[xiv]

The U.S. Navy named a destroyer escort commissioned in 1957 the USS *Ernest E. Evans* (DE-1023). *Evans* served four tours of duty at Yankee Station during the Vietnam War and was decommissioned in 1968. In November 2023 (Native American Heritage Month), the Secretary of the Navy announced that a future Arleigh Burke- Flight III class guided-missile destroyer would be named the USS *Ernest E. Evans* (DDG-141).[xv] Each sailor who will man the future USS *Ernest E.*

Evans will see the citation for their ship's namesake and perhaps draw inspiration to replicate his leadership and courage when our nation once again asks young men and women to sail their ship into harm's way.

Jim Tritten is a retired Navy carrier pilot who lives in a small New Mexico village with his Danish author/artist wife and an ever-changing number of cats.

CHAPTER 7

TO RUN OR NOT TO RUN?

BY TANYA WHITNEY

I've never been a runner. Matter of fact my mantra was why run when you can walk and still get there. Bicycles were for distance. The extent of my running for years was around the softball bases or shagging balls in the outfield. Therefore, when I decided to join the U.S. Army, I knew I had to do something about being able to run two miles. Especially since I hadn't played softball in a couple of years as I waited until I was almost twenty-two to enlist.

In 1983, there were not a lot of places nearby where I could run. I lived in a predominately rural area of farms and cow pastures. You might say why didn't you run in the cow pastures and my answer would be 'bull.' No not bull shit, which was also a concern, but actual bulls that were never happy when someone invaded their pastures.

So, that left the road where I lived. It had wide shoulders. I figured it would be okay. Boy did I have that wrong. At first, I ran with the traffic because that's what you are supposed to do. After the third time of being nearly run over because I couldn't see behind me, I switched sides. At least now I could see what was coming and duck into some-one's yard if necessary.

Wrong! Out running one day, minding my own business, an idiot decides to pass the car in front of him. Now mind you I always hugged the far side of the shoulder. Apparently, the passing truck decided to see how close he could get without hitting me. I felt the wind rush by my ear as the truck's mirror passed by. As the truck swerved back into the lane, I promptly turned around and went back home. I ran circles in the yard for the next few weeks before I left for basic training.

My basic training was at Fort Dix, New Jersey. For the first week, I don't remember running much. Afterward though, it seemed as if we ran everywhere. The worst were the unit physical training runs. I hated running in formation. The drill sergeant would always get mad because I would drop to the far back or on the side. One day he decided to make me the pacesetter on the next run.

As you can imagine that didn't go very well and only lasted one run. No one wanted to run at a fifteen-minute mile pace. One by one, the line behind me passed until I was once again at the back of the formation. I was happy. My fellow trainees were happy. The drill sergeant was not happy.

Eventually, I built enough endurance to run my two miles under the time limit. That pretty much stayed the course throughout the rest of my military career. The main reason? The older I got the less time I had to run because the Army, in their infinite wisdom, kept changing the standards. But I persevered until February of 2005 when I blew out my left knee.

At Fort Bliss, Texas for combat deployment training, I was coming off the back of a five-ton truck in full battle rattle when I landed wrong. I knew instantly something was very wrong as it hurt to even breathe. Diagnosis? I shredded my meniscus like a ripped and crumbled piece of paper torn from a wire notebook. The kind that never tears evenly and leaves little pieces of paper all over the place.

After surgery, I received the devastating news…I would officially never be able to run more than one hundred yards ever again. Now you may think how great that was. Nope! I had to do an alternate event. I couldn't swim (and still cannot) so that was out. The base I was assigned to was so small they did not have the bicycle required to test on. That left the two-and-a-half-mile walk.

For my age group at the time, I needed to walk the distance under thirty-six minutes. Now you may think, *oh that's easy, I can do that.* Well, guess what? It ain't! The first time I did the walk I barely passed. And to this day I fully believe the first sergeant took pity on me and stopped the watch early. The second time, six months later, I did a more respectable time of a little over thirty-four minutes. I timed myself so I knew for sure I passed.

Unfortunately, power walking did more damage to my knee than running. By this time, I had a little over two years to go for an immediate active duty retirement. I had over twenty years in the National Guard, but the medical evaluation board didn't give me much choice. Take a lump sum for a medical discharge under eighteen years of active duty but give up my Guard retirement. Or take the discharge with no money and wait twenty years to draw my Guard retirement.

Luckily, I had a colonel on my side who had no qualms about using the regulations to my advantage so that I could finish out my active duty time. For the next nine months, I stayed on extended profiles until I was locked in for retirement. By then I had changed age groups and had more time to walk. I took one more test, passed it, and went back on my medical profile until I retired.

I made it. No more running ever, no power walking, and no more worrying about it. I retired and returned home in time for my daughter's senior year of high school. Now you may ask why that little tidbit matters. Well, the irony of the story is that my daughter's cross country team needed a coach. I thought okay I can do this for the year so the team, especially so the seniors could finish out their high school year without missing their final year of running.

With limited knowledge of coaching and the inability to run, I figured it would be one and done. Also, as the previous coach was a teacher, I expected the Athletic Director would find another teacher to take on the team next season. So much for assumptions and expectations. I was asked to take over the team for the next season. I have been a high school cross country coach for over ten years. (I also assist with the school's track team as a pole vault coach but that's a story for another time.)

I took coaching courses and created a program that teaches teenagers how to run and enjoy it. I can't run distances with them, but I have my watch points that I shuffle to and from on the course. And for those who have trouble with distances, I rely on my power walk training from the Army to keep them moving. It's a little different from my days in the Army. Then I was learning how to run under a certain time. Now I teach kids to run for a better time. It doesn't matter if it's from forty minutes to thirty or eighteen to sixteen minutes. It's about setting goals and doing your best to achieve those goals. Something the Army taught me every day I ran.

Tanya Whitney, retired Army Master Sergeant, began writing as part of her PTSD therapy. While her writing primarily centers on her military service, she has also branched out in other themes. She is the author of two poetry books, *A Soldier's Journey Home,* and *A Journey to Healing,* along with her first novel, *Fatal Secrets.* Her award-winning poems and short stories have also been published in various anthologies in the U.S. and England.

CHAPTER 8

THE TEACUP

BY ROSALIE SPIELMAN

The toddler was finally asleep. The young mother's bare feet slapped against the military housing linoleum as she made her way through the small home. She was searching for her favorite teacup—one that Gran had given her, telling her to imagine a "Gran sugar-hug" when she used it. The saucer had been broken in the last move, so this teacup was her last link to Gran. It was irreplaceable.

She finally located it, sitting forgotten next to the computer from the last time she had checked to see if he had emailed. She stuck a fingertip into the tea, and finding it stone cold, headed for the kitchen. She stood in front of the microwave as it reheated her chamomile tea, rubbing her swollen belly and wondering if standing this close to the humming machine would hurt the baby.

When the microwave beeped, she took the dainty flowered teacup and began to wander through the house, muttering to herself. Her whisperings weren't the shopping list, but rather, a two-sided conversation with her husband. Since he was 8,000 miles away, she answered for him as he would have, and smiled at his sweet answers.

She continued the conversation as she weaved her way around the toddler toys strewn on the floor, and the sleeping toddler himself, but her path to the couch was interrupted by the sound of a car door

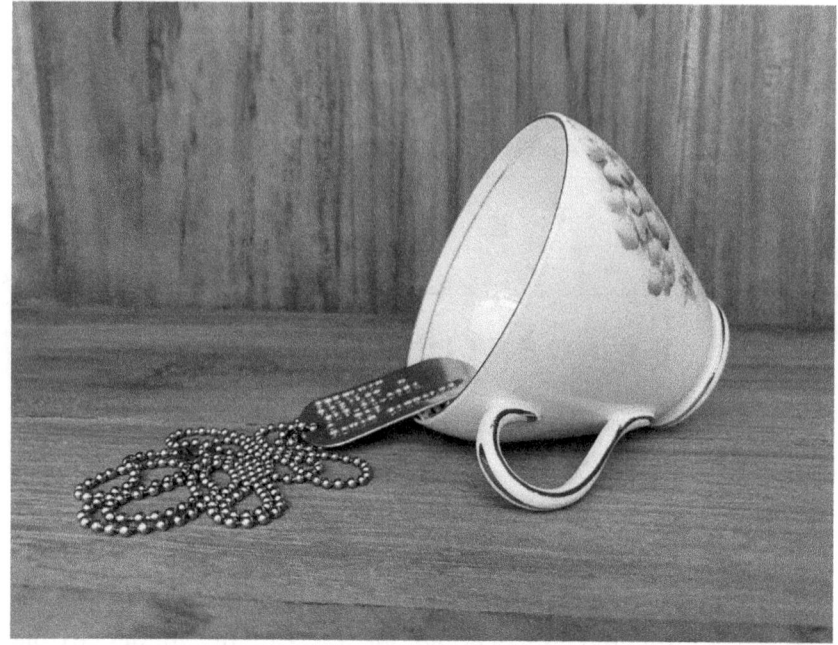

outside. Curious, she changed direction and went to the window to peek around the curtain.

She went silent, her words forgotten. A car had parked across the street, in front of her neighbor's house. Her heart squeezed as she watched a man in Class A's shut the driver's side door and join three other soldiers who were dressed the same. She held her breath as they stood in the middle of the street and glanced around. Her lungs burned as she watched to see which house they would turn to.

Which lives they would ruin.

She barely noticed that she was clutching the fragile teacup to her chest with both hands, little waves of tea breaching the edges of the fine china and splashing out onto her tank top as she shook.

The eyes of one of the soldiers locked onto hers. His green uniform had shiny silver on it, but she didn't know who he was. He didn't smile.

The man turned to the others, and her arms went slack, the remaining tea dribbling out onto the floor and splashing back up onto her bare ankle.

Then, as one, the soldiers turned to face her house. To face her.

The toddler woke with a start and began to wail when Gran's sugar-hug teacup shattered on the hard floor.

Rosalie Spielman is an award-winning author, mother, veteran, and retired military spouse. She was thrilled to discover she could make other people laugh with her writing and finds joy in giving people an escape from the real world. She is a member of Sisters in Crime and the Military Writers Society of America, the latter awarding her work with a gold medal in 2024. Rosalie strives to *provide you an escape...one page at a time.*

CHAPTER 9

GHOSTS OF YORKTOWN, SEPTEMBER 11, 2001

BY DARCY GUYANT

I stood in the middle of the Yorktown Battlefield, the site of the final major American Revolutionary War battle, which took place in September and October 1781. I was directly between the British and Colonial Army positions. This location would have been deafening and deadly, with canons and muskets spewing lead across the open field. But at the moment, it was eerily quiet and still.

It was 5:00 p.m. on the afternoon of September 11, 2001. I was on my way home from the U.S. Coast Guard Training Center (TraCen) Yorktown, where I served as the Assistant Chief of the National Search & Rescue School. The road to TraCen Yorktown passed directly through the Yorktown Battlefield, amidst the location of the fiercest fighting during that famous battle, the battle that essentially ended the war and led to our independence from British rule in 1783. I had the privilege of passing through these hallowed grounds daily as I commuted from my home near Williamsburg, VA, to the USCG TraCen.

It had been a surreal, confusing, and emotional day for me, as it had been for nearly everyone. I needed time alone to reflect on the day's events before going home. So, I stopped at an observation area on the battlefield, walked to the top of a defensive berm constructed by Colonial forces, and stood silently. There was not the slightest hint of wind. The sky above was clear and bright blue; no clouds or jet contrails marred the sky. The sky above was some of the busiest airspace in the country. There were always jets crisscrossing the sky. But not on this afternoon.

Earlier that morning, I watched two F-15 fighters from Langley Air Force Base pass directly over the TraCen, heading north. They were flying low and very fast, their afterburners roaring. But now, six hours later, nothing was in the air except for a few Air Force and Navy fighters, who remained vigilant for additional attacks.

Sparrows and robins were fluttering here and there, oblivious to the mandatory nationwide grounding. I was struck by the eerie quiet. Usually, an aircraft could be heard somewhere in the distance, operating from one of the many Army, Air Force, and Navy bases in this part of Virginia or from the busy international airport just miles away. But at that moment, there was nothing but quiet, except for birds singing in the distance, grasshopper wings clicking as they flew from spot to spot, and two deer munching on the grass nearby.

My great (x5) grandfather, Luke Guyant, had fought here, 220 years before, during the Battle of Yorktown. He was assigned to General George Washington's protection detail. As a 7th-generation "Son of the American Revolution," knowing that my great-grandfather had been right here, where the future of our country was altered, made the day's events even more surreal and poignant.

As I stood there, I thought about the progression of my emotions throughout the day—from my initial skepticism of the news to my confusion over reports of more planes crashing. As more news and video began streaming on the TV, I felt disbelief. This couldn't really be happening! Then, as I watched live, the first tower fell and disappeared in a cloud of dust. I was stunned. I stood watching, speechless, as were the others standing beside me. Then, the second tower fell.

The twenty students in our Search and Rescue Planners Course had been in the classroom all morning and had no clue what had happened. The School Chief somberly walked into the classroom and interrupted training. All instructors, staff, and students gathered in the classroom as he solemnly briefed us on the situation. The training was canceled for the remainder of the morning so everyone could call their units and loved ones.

Now, hours later, as I stood in that serene setting, surrounded by the surreal silence, I imagined the ghostly presence of those who perished here, securing our freedom so long ago. I asked the same question millions of others were asking that day, "Why God? Why have You allowed this to happen to so many good and innocent people?"

Regardless of how deep and firm a person's faith is, wondering how something of this magnitude fits into a loving God's omniscient existence and sovereign plan are natural questions. I already knew the answer to this question, but I still found it hard to fully grasp and understand, especially when directly impacted and so close to home.

When I got home, my two teenage kids, my wife, and I watched the evening news together. The reports, images, and videos were all about that day's barbaric attacks and the tragic aftermath. When the news finished, I turned off the TV so we could talk and share how we were doing. I told my kids I was sure this day would have huge, long-lasting impacts on our country, the world, and potentially each of us personally. At the time, I had no idea how accurate that prediction would be and how deep that impact would affect each of us.

My son and daughter would both serve in Afghanistan. My son, a Marine Corps infantryman, would engage in fierce fighting in the Helmand Province of Southern Afghanistan. Our daughter deployed years later as an Army doctor serving in the Emergency Room at the NATO hospital in Kabul, Afghanistan.

My wife and I spent much time on our knees praying for their safety. I thank God for answering our prayers. Both of our kids returned home physically unharmed. However, some wounds are unseen. Those wounds take much longer to heal completely, if ever, and may leave lifelong scars.

I wonder how my great-grandfather Luke and his fellow revolutionaries would have viewed the heart-rending events of September 11th. They had fought for our independence and the idea that a nation built on liberty and justice could endure and thrive. I think Grandpa Luke would have been proud to see his 8th-generation descendants stepping up to defend what he and others had fought to establish. They would have seen in us the same resolve and unyielding spirit that carried them through their fight for freedom, witnessing the ideals they bled for still standing.

Commander Darcy Guyant, U.S. Coast Guard (retired), served in the U.S. military for 25 years, flying helicopters in the U.S. Army and the U.S. Coast Guard. In 2022, Darcy published his first book, a memoir, and in 2023, he began writing and publishing children's books. The stories in his first children's book series, *"The Coast Guard Adventures of Dolph and Gwen,"* are based on actual search and rescue missions flown by Darcy in an HH-65 Dolphin helicopter. Darcy's love of storytelling and his exciting flight experiences have been a major source of inspiration for providing stories demonstrating courage in the face of adversity, teamwork, and trust in their teammates.

CHAPTER 10

BOOT CAMP - SAN DIEGO NAVAL TRAINING CENTER
SEPTEMBER - DECEMBER 1964

BY A MICHAEL HIBNER

My Civics teacher at McCurdy School (in Santa Cruz, New Mexico), Miss Elvira Townsend, once made a comment to the effect that she was educating kids that would be going off to war after they graduated. I wasn't sure what she meant, but after graduation I had to go down to the post office and get a draft card. Not too long after that I noticed that a lot of the guys in the area were joining the Navy or Air Force, or going to college, or getting married. Being a quick study, I figured out that these were all ploys to keep from being drafted into the Army. It also became apparent that those who were drafted were not coming back or, if they did, not necessarily in one piece.

I graduated in 1962 as a seventeen year old, probably the youngest in my class. So, I had a little time, but decided it was best to keep moving. I worked in Santa Fe for a while, then headed to Oregon and attended college for a quarter, then moved to Las Vegas and worked there for a few months. I knew I was on borrowed time, though, so in August of '64 I joined the Navy.

I was never a good student. I just didn't like being there. I mostly made Cs. Teachers told my parents that I was capable of doing better—I just didn't apply myself. And I never did homework. I was really good at taking tests, though, so I was listening, sort of, and reading. I think I read every book in the school library. So, when the first thing the Navy did when I joined was give me a test, I had no problem with it. In fact, I did so well—the highest they'd seen in a while—that they gave me the orders for the fifteen or so guys that joined up the same day.

I joined in New Mexico, so on the appropriate day I made my way to Albuquerque for physicals and tests. The Navy put me up at a hotel in downtown Albuquerque, maybe the Hilton, and the next day sent a cab for me. The cab took me to the Santa Fe Railway station. Our recruiter gathered us up, gave me the orders for everyone, and put us on a train. Next morning we pulled into the Los Angeles station and switched over to a train headed to San Diego. I took rollcall and found that one of us was missing.

I was supposed to call a number when we got to San Diego, but there were pool tables in the train station. So, we got lunch and shot a few games of pool. Then I called the number. Delaying for an hour or so was probably not the smartest thing I could have done, but I didn't get any repercussions from it. I did think the missing enlistee would be a problem, but I guess that happens often. Anyway, it wasn't held against me.

Shortly after I called, a sailor walked into the station and began screaming at us. We were called all sorts of derogatory names, lined up, marched outside, and put on a bus. The next stop was the San Diego Naval Training Center; home for the next three months. Still screaming at us, the sailor lined us up. In the next few hours we were stripped of our civilian clothes, shorn of our hair, given inoculations, uniforms, skivvies, duffle bags, shaving supplies, ditty bags, and everything else we would need; not necessarily in that order. We kept nothing that we came with—everything was boxed up and sent home.

When we had everything, we were marched to our barracks. We were grouped with guys from other towns and cities until we had the requisite number for a boot camp company. We were Company 492; our Company Commander was Chief Petty Officer LeFevre, and we were his "Boots." We now found out why the Navy calls it "Boot Camp." We were to be torn down and rebuilt from our boots up.

When we got to the barracks, we were assigned beds, mattresses, pillows, and lockers. We were taught how to stow all the new gear. We were taught how to make our beds. There was a place for everything, and everything had to be in the right place. And everything had a name that had to be learned. Our beds were racks, our sheets were fart sacks, etc.

Well after midnight, after everything was stowed, and the racks were "dressed," we finally got to go to bed, or "hit the rack."

"Reveille" was at 5:00 a.m., or 0500 hours in military time. After showers, muster, and chow, a very interesting event took place. We were queried about our previous military experience. I had none. But others had been in ROTC or military schools, and some had had previous experiences in boot camp. From this group, there weren't very many of them; our "recruit petty officers" were assigned. The one with the most experience had lied about his age and joined the Navy when he was sixteen. When his true age was discovered, he was booted out. When he joined the second time, he had enough experience to be assigned the position of "Recruit Chief Petty Officer" or RCPO. A few others were assigned, an RPO1, a Master at Arms. The shortest one was assigned as flag bearer, others I've forgotten. Then we were all lined up according to height. At 6'2" I was the fourth tallest. This was a good thing as the six tallest were assigned as squad leaders. I soon figured out that being a squad leader was better that not being a squad leader. All the others, according to height, were lined up behind the new squad leaders, and I was now responsible for the Boots lined up behind me.

For the rest of the day, we learned how to march—forward march, left face, right face, about face, halt, etc. We learned that you start with your left foot. Amazingly, some didn't know their left foot from their right. We spent the rest of that day and the rest of the week learning all sorts of stuff you needed to know to get through boot camp. I don't think much of it was useful after boot camp, though, when stationed on Treasure Island, Fleet Admiral Chester W. Nimitz died. I, along with about five hundred sailors, donned our dress blues and marched in his funeral procession.

The second week was more marching, but most of the time was taken up with taking tests. I think there were at least eight with each lasting at least a couple of hours. One test consisted of a series of beeps. You were supposed to determine which of the two beeps were higher in pitch. They all sounded the same to me. Other tests were for hearing and vision. Unfortunately, neither is my strong suit. There were also math tests and general knowledge tests. And, remember, tests were what got me through high school.

The third week was service week. During this week, most of the Boots were sent mess cooking. Mess cooks in the Navy peel potatoes, wash dishes, serve food, and clean up. It probably wasn't much fun. But I didn't have to do it. Remember my comment that being a squad leader was a good job to have? Well, it turned out that squad leaders were not assigned mess cooking during service week. I was assigned to the USS *Recruit*. The USS *Recruit* looks like a ship, kind of, but sits in the middle of a giant parking lot. It is a training aid for recruits. It had a Captain, an Exec, and several other permanent assigned personnel. It also had a "Runner" that took stuff from the *Recruit* to other places on the base. So instead of washing dishes and peeling potatoes, I spent the week "running" things such as mail, or supplies such as toilet paper. Oh, I had to "swab the deck," but it was really a lot of fun and way ahead of mess cooking.

The fourth week, the results of all the tests we had taken were posted on the bulletin board. I didn't really understand the ramifications of the tests, but two of the Boots did. These two had been appointed as Education Petty Officers and were responsible, sort of, for making sure everyone passed the weekly tests we took on stuff we learned in *The Bluejackets Manual* and the Uniform Code of Military Justice. So, when the test results came out, these two were bragging about their combined ARI/GCT scores. They both had an ARI/GCT combination of 134.

"We're tied for the highest in the Company," one said.

I asked them how they figured out their scores, and they explained it to me. I checked mine.

"Hmm, my score is 135." My stock in the Company went way up that day. I was not only one of the squad leaders. I was also assigned as one of the four "Cheaters."

Being a Cheater was an unofficial assignment but held a lot of prestige. It was the job of the Cheaters to make sure that the recruits passed the weekly tests. The way this worked was that the four Cheaters were to sit in the middle of the front row of the class room. Each Cheater was assigned a letter, from A to D, that corresponded to the four choices for each question in the multiple-choice tests. If the test started at 1300 hours, The Cheater with the right answer to the first question held his pencil a certain way until 1301. Then the Cheater with the right answer to question two held his pencil until 1302, and so on. Of course, this demanded that the recruits utilizing the Cheaters had to be able to keep track of the time and the question number, and if two Cheaters held their pencils in the designated position you were on your own.

After the fourth week, boot camp became repetitive-marching drills, studying for tests, taking tests. Repeat. There were a couple of interesting things that happened though. One day our Company Commander marched us over to a fence that separated the Navy boot camp from the Marine boot camp. He called a company halt, and we stood at ease. After a while, a Marine company marched past us at double time, at least.

"Be glad you're in the Navy and not a Marine," he said. I had to agree with him.

Another day the Company Commander told me to report to a certain building and office. I did so. I was directed to an office where two officers were waiting for me. Because of my high ARI/GCT score, they were interested in talking to me about going to Officers Candidate School after boot camp.

"Sure," I said, "but have you checked my eye sight?"

They did.

"Report back to your company."

Two days later, I was asked if I was interested in Pilot Training School. Same question, same review, same result. They just weren't too interested in someone with 20/30 vision in his good eye.

Everyone was sent, one at a time, to discuss their future job in the Navy. Every sailor has to "strike" for a rating. I really wanted to become a Seabee. Seabees build bridges, roads, air strips and all sorts of things for bases and camps, sometimes in a war zone. But you must have prior experience in construction, so I couldn't strike for any Seabee rating. I had to make another choice. I didn't have any others, so I just wrote down three ratings—I don't remember what they were.

"Add ET to your list," I was then told.

"What's that?" I asked. I was told it was Electronics Technician.

"Okay," I said, and wrote it down. Of course that's what I got. I think you had to have an ARI/GCT score of at least 125 to qualify for ET A School.

One other situation that came up left me with a bad feeling. One of the squads had a recruit that for some reason would not take a daily shower. The Navy insists on cleanliness and daily showers are mandatory. So, the squad leader of this recruit organized a GI shower for him. This consists of the rest of the squad members placing the individual in a trash can full of soapy water and scrubbing him down with rough bristle brushes that were normally used for scrubbing floors. Though I understand the necessity for this action, I felt badly for the guy.

I don't think he made it through boot camp. Several didn't. If you couldn't pass the tests, you could be separated or discharged. If you got in trouble, same thing. One Boot found out he was allergic to wool. Since dress blue uniforms are wool, medical discharge.

About the fourth or fifth week, we marched down to the swimming pool. Now, you would think that a kid that joined the Navy would know how to swim. I didn't. And I wasn't alone. First, we were lined up at the high dive platform. We each had to climb up the ladder and jump off the platform. I've never had a problem with jumping off high things into water, even though I couldn't swim. I really can't explain that. So, in I jumped, dog paddled to the side of the pool and got out. One of the Boots froze when he got to the edge of the platform. He wouldn't jump. While one of the Company Commanders yelled at him, another climbed the ladder, walked up behind him, and shoved him off.

Next, we were instructed to jump in the water, take our dungaree pants off, tie the legs in a knot, then flip them over our heads to fill them with air and float on them. This does work, as a sailor managed to do just that when he fell off the catwalks on the USS *Kearsarge* (CVS-33) one night at sea.

But I couldn't do it. I couldn't fill the dungarees with air, couldn't tie knots in the legs, couldn't even get them off. I did get them unbuttoned, but that was it. Fortunately, you don't have to be able to do it to get out of boot camp. But you do have to be able to swim from the shallow end of the pool to the deep end, swim along the deep end, and then swim back to the shallow end. I managed to get to the deep end, across the pool, and back toward the shallow end, but I was floundering. I wasn't going to make it. But as I started sinking, I noticed that I could touch the bottom with my toes. So, I walked, with my arms thrashing the water, to the finish line.

Boot camp is all about repetition. The more you do something, the better you do it. So marching, calisthenics, studying, taking tests, inspections, was what we did, day in and day out. There are two ways to do everything: the right way and the Navy way.

Dressing your rack is a good example. The fart sacks had to be so tight that a quarter would bounce on it. Your locker had a place for everything, and everything had to be in the right place. Our uniforms had to be worn in the correct way. We probably had an inspection daily, either our racks, lockers, or dress. Our belt buckles had to be shiny bright. Our shoes had to have a spit shine. Boots that could spit shine shoes to a high luster could make good money doing that for Boots that couldn't, though there wasn't much use for money.

One fateful day, the Company Commander called a surprise inspection. We had about five minutes to get ready. I couldn't find my belt buckle. It had a specific place in my locker, but it wasn't there. I thought I might get away with not wearing it as our jumpers were worn outside our trousers with a webbed belt around our waists. The inspector did not check everyone's belt for a buckle, but he did check mine. No buckle, demerits. I think he must have seen the fear in my eyes. After inspection, I checked my locker. My buckle was where it was supposed to be.

But most everything else went well for me. I kept my job as a Cheater all the way through boot camp, but more importantly, I kept my job as a Squad Leader. Most didn't. If there was a problem with the Boots in your squad, it was your fault, and you could be replaced. But I had good guys, and I worked with them whenever we had down time. I held study time (mandatory), so that they had a better chance to pass the tests. You needed to keep the same squad members if you could, because when Boots left the company, the squads had to be evened out. So, if you lost someone in your squad, you got the worst Boot from another squad. Of the original squad leaders, only two of us weren't replaced.

All good things must come to an end, even boot camp. Our last day consisted of preparing to leave San Diego NTC. We sewed our Seaman Apprentice stripes on our uniforms, packed all our gear in our duffle bags, went to disbursing and drew our pay and received our tickets home and to our next duty station. And orders. I'd found out about a week earlier what rating I'd be striking for. I would be going to ET-A School on Treasure Island.

When the bus pulled up, I picked up my duffle bag and ran for it. I was the first one on board.

A Michael Hibner - As a Navy Electronics Technician-Radio (ETN2), Arthur Michael Hibner spent the first eighteen months of his enlistment in California in Boot Camp, and A and C-Schools, then thirty months on board USS *Kearsarge* (CVS-33), mostly on the South China Sea/Yankee Station off the coast of Vietnam. Thus, he is a charter member of the Tonkin Gulf Yacht Club. He now writes historical novels and poems.

CHAPTER 11

NORMANDY

BY RICHARD EPSTEIN

They fell like dried leaves in a strong autumn wind.
They fell and they fell until their branches were bare.
Jerries tried to keep them there.

Juno, Gold, Omaha, Utah and Sword. June 6, 1944.
Almost seven thousand vessels made their way
through a million mines, tank turrets and bunkers
along the high ground of the fortified Wall.
Six hundred and thirty Tommies killed by U-boats
as they practiced off the English coast.
Twenty-four thousand paratroopers the night before
and one hundred and sixty thousand men
brought by sea landed on the beaches of Normandy.
No one knows how many fell that day.

They fell like dried leaves in a strong autumn wind.
They fell and they fell until their branches were bare.

Richard Epstein, a long-time resident of the Washington, DC area, has been a featured reader at the U.S. Navy Memorial, The Vietnam Woman's Memorial, the Orange Bear (New York City), and others. He is the editor of two veteran anthologies, and his poetry has appeared in *O-Dark-Thirty, DEROS, Incoming, A Common Bond,* and *Schuylkill Valley Journal.* For twenty-nine years Richard hosted an open mic venue each Memorial Day and Veterans Day on the National Mall adjacent to the Vietnam Veterans Memorial and via Zoom.

CHAPTER 12

FROM EUROPE, WITH LOVE

BY CAROL MARKOWITZ

Long forgotten, tied in bundles with red ribbon, letters from long ago. They're filled with the hopes and dreams of young men and women. Filled with promises and expressions of love and longing. Most of the writers are gone now. But the letters remain in their envelopes, waiting to be discovered. I found packets of them. Carefully preserved, kept in order. Dates starting in December 1942. It was long ago, a different world. When I read them, they played out like a romance novel. Who would make it home? Who would not? Would they find love? Would they marry?

Letters from WWII were a lifeline for soldiers and sailors. The only link to home and family. The connection to sweethearts, mothers, and fathers, who waited anxiously for each one. On the battlefields of Europe and the vast Pacific Ocean, young men waited for mail call. The precious letters were saved, read, and read again. They asked for football scores, who's dating whom? How's the family? Some were little V-Mail letters and some were airmail. Some carried trinkets like embroidered handkerchiefs, hand drawn greetings, or little black and white photographs.

When I opened the packets and started reading the letters I became fascinated by these soldiers' lives. They were men I never knew, writing to my mother. All of them gone, there was no chance to ask them how they felt or what they meant. I had only one side of the correspondence. My mother kept all their letters, but her letters were lost. One set of letters started in December 1942. He included pictures. He was at home with his family, the other young men must have been his brothers. They were horsing around with him. His uniform looked neat and new. He was a newly minted soldier, a medic assigned to an Evac Hospital. He was tall and slim, with a handsome face, dark hair, his eyes looking directly at the camera. He wrote from his first posting in North Africa.

"How's you all? Guess where yours truly is - I'm in dear old Africa- and it's quite the place. I can't tell you much about the place as yet. Matter of fact I can't tell you anything...I could sure use some of your mother's lemon pie- hmmm- give her a big kiss for me and you can have a few for yourself. But!! Be careful who gives them to you-"

His letters were full of jokes and teasing, always upbeat and funny. There were times I found myself laughing out loud at something he wrote. Then I would remember these words were written more than eighty years ago.

They met at a dance at Camp Kilmer. He was stationed there until he shipped out with his hospital, she was a volunteer. Dubbed "Kilmer Sweethearts" the volunteers' were to serve cookies and punch and dance with the young soldiers. They were often surrogates for girls left at home, someone to talk to and laugh with. Edward "Mac" McIntire met my mother, and they danced the night away. After that night they spent every free minute together until the inevitable happened; he shipped out. It seemed that they were very much in love.

I followed his story, reading his words, often reading between the lines. Trying to get to know him. I became a detective; I found the history of the hospital and some oral histories of other men and women in the unit. His story was becoming clearer and more complete. His personality came through in his writing. *"What were you trying to do? Rub it in? You go see Woody Herman and I listen to Arab drummers - you know even Woody could learn a few hot licks from these boys."*

The hospital was stationed in Casablanca long enough for the men and women to begin to feel at home there. They traveled to Rabat to enjoy the beautiful beaches and swim in the ocean. Restaurants in the city served fresh seafood, a break from Army chow.

He wrote almost every day. He was often flirty and romantic. *"Are you joining the WAACs? Good deal; I'll see if I can pick up another camel then we can go for moonlight rides."*

The fighting ended in North Africa; the Allies were victorious. The hospital was designed to be mobile, and they began to follow the front. They had set up tents in Casablanca and now they packed up and headed east. They traveled by train to Tunisia, from there by LST to Sicily where they were needed badly. After the Allied invasion, the hospitals there were overwhelmed with patients. In addition to battle casualties, they treated malaria, Dengue fever, and Malta fever. He wrote about his *"all-expense paid moonlight Mediterranean cruise"* and setting up the hospital in Palermo, this time in partially bombed out buildings.

He wrote from Palermo. *"Your globetrotter brings you news from the land of plenty – a mess of mountains surrounded by mountains, water, and C rations. The Malaria bug is about the size of a P-38 and twice as fast. You take one to bed with you, and you don't have a gun, you come out second best."*

After reading dozens of letters and following the progress of the medic and the hospital, I began to feel that I knew him, his manner and outrageous sense of humor. As I carefully took each letter out of its envelope, I felt a real connection to the past. Some of the letters were V-Mail, I struggled at times to read the tiny print. Some were airmail, the paper thin and brittle. He wrote in a flowing script, always using a fountain pen. Sometimes he cursed the pen if he left inkblots on the letter. He talked a lot about his friends too and I was getting to know them as well. I began to feel I had to tell their stories. It was the beginning of my journey, becoming a writer. The journey took me deeply into the history of WWII.

"Many things have happened since you last heard from me. The main thing being the losing of my two best friends. Yes, the boys finally made it; Will and John left yesterday for a crack at the Rangers. It's a rough life but then they are pretty rugged guys, so I guess they'll make out...We miss them like the devil...Well Old Bean it's pretty late and we have a jug of 15-year-old brandy to kill yet...If you can't read this blame Hitler, it's a German pen."

His friends were fighting the Wehrmacht with Darby's Rangers. The Rangers had joined the fight in the Italian mountains near the village of San Pietro. Mac was right, it was a rough time for the men, they were mud-caked, unshaven, cold, and exhausted. The fighting went on for weeks, often at close range. Their boots were falling apart from the wet freezing weather, and many had badly swollen feet. They were continuing to push the enemy back. When the villagers started to emerge from caves in the mountains after the Allied victory, they found their homes destroyed. Their crops and vines were ruined by the intense fighting. All the trees were cut down by artillery fire. Men and vehicles sank in the mud. The mountainous country was cold and desolate. But Will and John would find their greatest test was yet to come.

Mac was a dental technician, and he sometimes wrote about his day at work in the hospital. *"Had a bad day from the start. Practically froze at roll call. Spilled coffee all over me at breakfast and cut the hell out of my face with my razor. So began my imperfect day. When I went to work, things really did start to pop – Blew the sterilizer up the first crack out of the bag. Lost my temper. This was a bad thing because I had two dentures to finish. One of said plates got caught in the rag wheel and sailed across the room; the other (I was mad by now) I threw the damn thing out the window. There was more but I'm out of space."*

On New Year's Eve he wrote. *"This is my last letter. That is to say my last letter in the year 1943. May '44 break the mailman's back. Well old bean I'm off to get stinkin' like my friend Lincoln."*

With a few days' leave on Sicily, Mac wrote about a tour of the island. *"Your globe trotter has just returned from vacation, and quite a trip it was. We traveled over the whole island stopping where we liked. You would never think a war was going on not too many miles away. Spent the better part of two days at Mt. Etna at a small town, Taormina by name. A prettier place one could never hope to find."* He enclosed black and white photos of the village with Mt. Etna in the background. He told her if he could, he would move into one of the little houses there, he liked it so much. She kept those photos too, pasted in a scrapbook.

He wrote about his friends; he hadn't heard from them, and he was getting worried. There was bad news from the front. *"I heard it's pretty hot up there and I don't mean the weather."* As I read, I was worried too. I felt like I knew these men and I was afraid for them. I started to read more of the history.

The Rangers' next mission was to capture the village of Cisterna. They were to take the main highways leading from Cassino to Rome. Infiltrating the German lines at night and then storming the town itself. It would be a difficult mission, but the men felt confident in their ability to defeat the Germans once again. When the 1st and 3rd Ranger Battalions moved up to take the village, they came face to face with a German armored division, infantry, and a parachute division. Will and John were in the 3rd Battalion.

The Rangers, traveling light and carrying only rifles were outnumbered and out gunned. There was little they could do against German tanks and heavy artillery. Some of the men tried to stop the tanks with grenades, blowing up the tank and themselves with it. They took cover and fired continuously but with little effect. They had been cut off, and the Germans were advancing on them. Over seven hundred men marched into Cisterna, only a handful made it back to the American lines. The rest were killed or captured. I read this and just like their friends 80 years ago, I was hoping for the best but fearing the worst. Weeks later Mac heard from John. He escaped by taking cover in a deep drainage ditch. It was a relief to hear that John made it back, but Will's fate was not known. John wrote, *"only guts and luck got us out, and mostly luck."* I was still worried about Will, Mac's joking, happy-go-lucky friend. Eventually they got word, Will was captured and was a prisoner in a German stalag. I was relieved, and I can only imagine how his friends and family felt when they got the news.

The hospital continued to follow the front. After Sicily they were on the Italian mainland, then on to the beachhead at Anzio. They followed the 7th Army to the south of France and then north toward the German border. They were treating patients in the Vosges Mountains in one of the coldest winters on record for Northern Europe. Equipment and gas tanks froze, soldiers were treated for frost bite as well as battle injuries. They were treating terrible casualties from the fighting there. Finally, they crossed the German border. Mac described it as being *"on the wrong side of the tracks, if you take my meaning."* The hospital followed the Army toward Munich and what would be one of their greatest challenges, the Dachau Concentration Camp.

As I kept reading, I saw a change in the tone of his letters. He talked about the possibility, even after all the years spent in Europe, he would be reassigned to the Pacific. *"Not much to report at this point, just a lot of blown-up towns and a lot of sad Germans."* Mac was only 26 years old in 1945. He seldom revealed his feelings or emotions, his letters always peppered with jokes making light of terrible situations. His "midnight cruise" to Sicily was described by others as a frightening time; their supply ship sunk by a German bomber. He was sick with malaria at Anzio. Threatened by German troops, the hospital retreated from the German border during the Battle of the Bulge. They treated terrible head wounds from "tree bursts" in the French mountains. His best friend described the slaughter at Cisterna to him. Those experiences took a toll. Like his comrades, he had come to hate the Germans. The young carefree soldier of 1943 seemed more hardened and cynical. He was not assigned to the typhus wards at Dachau, but like all the soldiers, he was ordered to tour the camp. General Eisenhower made sure there were many witnesses to the horror.

"Today is Mother's Day in Germany. Doesn't mean a lot does it. All the little mothers will be out with their pappa-less children. Went up to Dachau, another Hitler Horror Camp, yesterday. Not a very pretty thing to see; but one that you're not likely to forget. We saw piles of bodies that had been stacked up like so much firewood. For the most part I guess they starved to death – not a very pretty picture. The only good part is seeing the SS troopers take care of the messy jobs around the camp. The prisoners really give them a bad time. For the most part the SS boys just go around with their heads down murmuring 'Alles kaputt.' So much for my day at Dachau. Just getting in condition for the Pacific. Don't say it!"

He enclosed a "Sad Sack" cartoon in the envelope. A soldier is pictured dreaming of going home and getting married. Then he sees his sergeant with travel orders for the Pacific. Mac wrote his own name in the cartoon.

If you're wondering how it all ended, Mac and my mother never had the chance to "paint the town red." Mac went home to California and married a local girl. My mother married too, a sailor home from the Pacific. They never saw each other again. Mac's buddies, Will and John, made it home to their families. My mother never talked about Mac; she never really talked about the war years at all. It was only after her death I found her letters tucked away in a box, carefully kept in their envelopes. My mother loved history; she spent her career teaching history in high school. Those letters represented an important part of American history, but I don't think that was why she kept them for all those years. I believe she kept them because they were personal treasures. They represented cherished memories of the war years.

The letters were of a time that was sad and hard in many ways. But for the men and women who lived it, that time would always be special. The war brought horrors and fear and death, for many the darkest days they would ever know. In profound irony, sometimes difficult to reconcile, these were also their best years. It was a time when they were part of something greater than themselves. Something with a noble purpose which they would never experience again.

And there were memories of young love. Love that blossomed over thousands of miles. In letters written in tents and bombed out buildings in the heat of summer and winter's cold. Men living in mud and snow in the hard reality of war. Love that was so passionate and for some so fleeting they would never experience anything quite like it again.

These were the experiences of World War II. The veterans are growing old and dying every day; soon they will all be gone. But the memories of their experiences will live on in their oral histories, journals, and their letters. Through these we can know them and see the world through their eyes. It was a different time; the war touched every life in America. Ernie Pyle was thinking about it on a troop ship bound for Sicily in 1943. *"In the heaving darkness of the Mediterranean I realized how everybody in America had changed... Everything in this world stopped except war and we were all men of a new profession out in a strange night caring for each other."*

May they rest with honor, and in peace.

Carol Markowitz is a retired educator, having spent more than forty years providing services to children and adults with autism. When Carol discovered an extraordinary packet of letters in her late mother's belongings, she knew she wanted to write about them. Carol is a member of the Military Writer's Society of America, the Author's Guild, the U.S. Naval Institute, and the National WWII Museum. Her story "Missing in Action" was published in *Snapshots: An Anthology* in 2023.

CHAPTER 13

REFLECTIONS: SERVICE TO OUR COUNTRY

BY JAMES BULTEMA

When you're nineteen, you don't think about dying; your thoughts and actions are in the moment - for now, this day, this minute. For most teenagers, that perception never changes, but for those of us who choose to serve our country, our take was much different. My transformation came on an early morning in Dong Ha, Vietnam. I was knocked out of bed (I know, all you Marines and Army dudes are crying, out of *bed* - what?) by massive explosions landing all around my hut. It was 1968, and my first thought was F***, I'm going to die today. Then the teenager in me said, "That's not fair; I leave for home in three days - *three days.*" I immediately crawled under my bed and started praying. "God, please let me live. I want to go home and live my life." I was a guy who didn't spend much time praying; it just goes to show who you turn to when you believe you are about to eat it.

I guess The Man couldn't decide, as rockets kept pounding our area. I could feel the concussion from the explosions shake my body. You feel so helpless lying there, unable to fight back. It's a moment in which you know you will live or die. I asked myself one final question, "What were you thinking when you volunteered for this?" I expected this was the end. But don't read anything into that; I wasn't ready to check out. Hell no. I had my entire life before me.

Well, I guess I gave the ending away, as here I sit all these years later, writing about this absolute life-or-death experience that I have kept chiefly tucked away for over five decades—meaning I don't talk about it much. So why write about it now, you ask—because not more than fifteen minutes ago, while having my morning cup of coffee, I was reading an article in *Dispatches*, a magazine published by the Military Writers Society of America, which was asking for submissions for an anthology. The title was "Reflections: Service to our Country." For whatever reason, as an author and member of this fine organization, it struck a note with me, and I began smacking away at my keyboard, writing about this life-altering experience of survival. So many can't. Just writing this brings tears to my eyes. These memories are tough to write about. I suppose that is why I have never written anything about them. It's just easier to live your life...like I prayed for. It's now ten minutes later, and here I sit, still unable to write a word as my emotions are flooding out, blinding me. I had no idea these feelings were so strong. As hard as this is, perhaps it's a good idea to finish this and cleanse myself of this ugliness called war.

I suppose my decision to serve my country and that moment under my bunk started with my dad and uncle, both war veterans who had 'II' behind it. Like me, I'm sure my father was a proud veteran— but he seldom talked about his war experiences except once. I don't even remember what we were doing that moment he decided to talk, but he did.

Herman "Duke" Bultema was in the 8th United States Army Air Force, stationed somewhere in India. I'm a genealogy freak, and my research suggested it was probably Panagarh Airfield in West Bengal. He said he was a crew chief on board the B-29 Superfortress bomber.

I recall he pulled out a stack of photographs, and there he was, this young guy in uniform staring into the camera.

What I remember most of that day is him telling me of his moment of truth about life and death. And like his son, it stayed with him all those years. He said he was at an airfield in Burma, as I recall, when Japanese bombers attacked. He told how a massive bomb shook the ground, not because it exploded, but because it was a dud. He didn't elaborate about his thoughts that day, how he could easily have been killed, but that was his moment of survival.

My uncle, William "Bill" Wiebenga, was the polar opposite. He often told me of his experiences during the war and was so proud that he made it a point to design his headstone with an inscription about his service to his country. He didn't want anyone to forget. His moment came in Guadalcanal when the Army fought alongside the Marines to take the island. It was the first major Allied offensive in the Pacific Theater. He was shot in the upper chest during the battle. Now, he didn't tell me what he was thinking as he lay there on the jungle floor with a bullet buried in his body, but I can only imagine. What he was most proud of about being shot was that the doctors were never able to extract the bullet and that, to his dying day, it was his daily reminder of his war.

He was so proud of his service. I'm sure that, between my dad and uncle and growing up with John Wayne movies, the patriotism they ingrained in me made me glow with my determination to fight for my country. When my war started, I was gung-ho ready to go. While many of my generation fought against anything to do with Vietnam, especially volunteering to serve, I was worried the war would end before I got there.

After boot camp in 1966, I was stationed at Pease Air Force Base in New Hampshire. Being in the Air Police, my job was "humping" B-52 nuclear-loaded bombers. I recall carrying my ancient M-2 Carbine around my assigned aircraft, guarding against any possible enemy invasion. We didn't call them terrorists in those days. But an invasion never happened, and the opportunity finally came when I could volunteer to go to Vietnam. I couldn't wait. Let me at those commie bastards.

That first breath of air I took coming off a civilian airliner in Da Nang was like something I had never experienced before, having come from the shores of Lake Michigan. The air was thick and damp, carrying the salty tang of the South China Sea mixed with the acrid smell of jet fuel and the earthy musk of tropical vegetation. It marked the start of something I couldn't yet comprehend but would never forget. Just a few steps down the ramp, the sound of F-4 Phantoms screaming overhead bounced off my body. The activity on the tarmac was mind-boggling. Jeeps, fuel trucks, airmen, and aircraft taxiing reminded me that this was war and that I was no longer in the U.S. of A.

I was assigned to perimeter security, wearing my blue plastic helmet with regulation bloused pants. My job was to check Vietnamese coming and going from the base. I was warned to look out for the Viet Cong because they looked just like their peaceful brothers from the city, except the former wanted us dead and gone. My assignment was boring; I didn't find any Viet Cong and turned to counting how many flies I could kill in a shift. This was BS. I came here to fight communists, not be a gatekeeper.

I heard at headquarters that the Air Force needed some Air Police volunteers for Dong Ha. I was told it was the home of the 3rd Marine Division, responsible for the defense and operations near the Demilitarized Zone. The base was located just seven miles from the DMZ. "And oh," said the second lieutenant briefing me, "the base is regularly hit with artillery and rocket attacks due to its proximity to North Vietnamese Army positions."

Perfect, I thought, finally action. I'm sure my dad and uncle had something to do with that decision. Flying into the Marine base in an Air Force C-130 was a wake-up call. There I was, standing at the

(Photo courtesy of the author)

rear of the aircraft with my duffle bag, and the crew told me that they would slow down but not stop because the enemy loved to hit the runway with rockets whenever the 130s showed up. As we circled the base, I saw proof. There, pushed to the side, was the burnt-out hulk of a C-130. We hit the dirt runway with a huge thump. Seconds later, I was told to jump. I threw my bag first, and then I went right after it. I remember rolling a couple of times as I hit the ground and hearing the engines roar from the C-130 tearing down the runway to get airborne. Welcome to Dong Ha.

Just days after I arrived, our large ammo dump was hit in the course of the NVA's offensive operations during the Tet Offensive's later phases.

All sorts of munitions flew through the air as massive explosions continued all day. I told myself you'd better watch what you wish for.

While in the Air Force, which some called the gentleman's service, I gravitated toward the Marines, whom I greatly respected. I met many of them at our "club" called the Cow Palace Lounge, a large bunker where you could have a beer. As I write this, I'm looking at my souvenir mug next to my Zippo lighter with my nickname "Bull" placed above a nicely inscribed map of Vietnam. At the Cow Palace, I met two Marines with whom I became fast friends (sadly, I can't remember their names). One was highly outgoing, but his buddy was quiet and didn't say much. I listened closely as they told me of their life out in the bush, involved in firefights with the NVA and moving all around I Corps. What a difference from my world, sitting in a perimeter bunker and guarding the Air Force's radar site. I was told the Marines would never lower themselves to do that job.

I saw them a couple of times over the next months, and on each occasion, they would tell me their war stories. We became good friends—what happens in the military when you experience war together. One day, I was assigned the front gate, and my talkative buddy came up and said hi. My first question was where was his partner? The answer haunts me to this day.

They were in a helicopter, he told me, when it crashed, settling at an angle. As the Marines jumped from the bird, our quiet friend was struck by a still-rotating blade, striking him in the head and killing him instantly. That moment shocked me because I rarely heard of casualties, and I don't recall losing anyone in my unit. The loss of such a close friend felt deeply personal. In my life, whenever the war comes up in conversation, my thoughts immediately go to him—how he gave his life for our country and was never granted the chance to live the life he deserved, just like the more than fifty-eight thousand others who made the ultimate sacrifice.

After my one year in-country I was ready to go home. Waiting in Da Nang in this large holding area, where I remember they were blasting the latest tunes, a song came on that soon had us all singing along, like a giant karaoke party. It was The Animals' "We Gotta Get Out of This Place." When the chorus came up, it was deafening.

As our civilian airliner packed with nothing but GIs lifted off from Vietnam, the cabin erupted with one of the loudest cheers I have ever heard. It was one of the happiest moments of my life, leaving the war behind and heading home in one piece. What I didn't realize at that moment is that you can never truly leave a war behind; it clings to you like a shadow that never fades.

After landing in San Francisco in full uniform, I was heading to my next gate when several "long hairs" started yelling "baby killer" at me—and other insults I have managed to cleanse from my memory. I couldn't believe it. Here I was, willing to give up my life for these people, and I got called a baby killer. That remains one of the most repulsive experiences of my life. I think of all the men and women who gave their lives for this country and how that was lost on so many Americans during that unpopular war. Today, whenever I see a Vietnam vet, I say, "Welcome home, brother" or "Welcome home, sister." We share a bond like no other—a connection forged in sacrifice, resilience, and an unspoken understanding of what it means to serve and endure.

As I reflect on those who came before me and those who will follow, I'm reminded that the spirit of military service is what keeps our nation strong. It's a call to honor not only our collective sacrifices but also the values our military upholds, ensuring that this legacy is never forgotten.

Welcome home.

James Bultema is a military combat veteran and a retired LAPD officer with over twenty-five years of experience on the streets of Los Angeles. His debut novel, *Sea of Red*, became a bestseller, earning six prestigious awards and recognition as one of the top military thrillers of 2023. *Red Lines* became a #1 Amazon Bestseller on its release. With a degree in history and a passion for thorough research, Bultema ensures his depictions of modern warfare and cutting-edge weaponry are as accurate as they are compelling. His riveting stories, full of unforgettable characters, resonate long after turning the final page. His new book, *Invaders of the Homeland*, a police procedural, will be released later in 2025.

CHAPTER 14

NIGHT REFUELING

BY HARRY STEVENSON

Aerial refueling a fighter is a challenge at the best of times. Mentally and physically demanding, it was as sure a way to tell the men from the boys as you could find in a fighter squadron of the 1970s, outside of actual combat. Night refueling was yet a better separator.

My aerospace vehicle was the F-4C Phantom II. The Phantom could take off at 55,000 pounds and carry lots of "stuff" under its wings. The Phantom had two crew members: the pilot and a Weapons Systems Officer, known as the "WSO" or "GIB" (Guy In Back). Operating the F-4 to its maximum required a team effort.

F-4 Day Refueling (Photo courtesy of the author)

Imagine trying to fly your 20-ton fighter, roughly the size of a World War II B-25, behind and below a massive flying gas station (KC-135). The object of this insanity: maintain a position so that an airman, the tanker's boom operator, or "Boomer," could extend a probe on a pipe, and plug the nozzle into your refueling receptacle located just aft of the rear cockpit. Fun at 20+ thousand feet and 300 knots! Once in position, you must stay there, no matter what: clouds, rain, or turbulence. If the tanker turns, you remain in position and go with it.

The refueling window is a six-foot virtual cube in space. The boom has a green circle, called the "apple," painted in the middle of the boom extension for visual reference. Your goal is to get the boom extension on the green apple and stay there. This is especially difficult for the F-4 pilot because he can't see it, except in his rearview mirrors. The only feedback an F-4 pilot has is the director lights on the tanker's belly and the calm soothing voice of his Weapons Systems Officer.

It was mid-1975, and I was Captain Steve Stevenson, in the 613th Tactical Fighter Squadron at Torrejon Airbase outside Madrid, Spain. I had been flying the F-4 for three years and was the Flight Commander of "C Flight" in the 613th for over a year. I had five F-4 crews assigned to the "C Flight" and had been a Flight Lead for nine months. The crews I normally flew with were from my flight, and I was responsible for their supervision, training, and performance.

That night, I led a 2-ship of F-4s, call sign "Duddy 51," down to the Malaga refueling track in the middle of the Mediterranean for night refueling. Then, each aircraft would fly a night low-level back north across Spain, with a simulated special weapons delivery on targets in the middle of Spain.

My WSO that night was First Lieutenant Jay Kittle, a returnee from the Southeast Asia air wars. Jay was a superb backseater, great with the radar, and good at flying the F-4, when he was offered the chance. We had flown together on numerous missions, and I enjoyed flying with Jay.

My Number Two was First Lieutenant Robert Parker whose grade books from F-4 initial training and our squadron checkout indicated occasional problems with aerial refueling. His WSO was First Lieutenant Johnny Wyatt.

A note on Parker's WSO, Johnny Wyatt: Johnny came to our squadron from the Southeast Asia war. While Johnny was an excellent WSO, most of the jocks initially saw him as a Texas ranch boy having fun flying in jets. That changed after a March Wing Officers Call where Johnny received the ninth and tenth Oak Leaf Clusters to the Air Force Distinguished Flying Cross earned in Southeast Asia.

Stories were circulating that Johnny was on a mission that used laser-guided bombs to finally drop the Paul Duomer Bridge northeast of downtown Hanoi, Vietnam, the toughest, most resilient, most heavily defended target in North Vietnam. We had lost more fighters trying to take out that bridge than any other target in North Vietnam. Johnny was **not** just a Texas cowboy!

OK, briefing, start engines, taxi, and the takeoff all went well. Two smoothly rejoined and we were soon climbing south from Madrid toward Malaga and our KC-135 tanker, callsign "Dobby 5-4." The Malaga refueling track ran straight down the middle of the Mediterranean. Joining the track just off Malaga, to the north, we could see the resort lights of towns along the Spanish coast, to the south, we saw the lights of the North African coast. But as we progressed east down the Mediterranean, the lights faded. Soon there were no lights to be seen, even the stars disappeared in the clouds that arrived.

We established radar and radio contact with our tanker. Jay had him on the radar before I could talk to him on the radio. The tanker was on the east end of the track and had just turned west toward us. With minimal vectoring, at the appropriate point, Jay radioed "Dobby, start your turn!" We had been closing on each other at 700 mph. The tanker rolled out of his 180-degree turn, and we finished co-speed and 1,000 feet below him. Perfect!

The plan for the night called for a 4,000-pound fuel transfer for each aircraft. First up, my aircraft would take 3,000 pounds of JP-4 fuel and then move to a position just off the tanker's left wing. Two would then drop off the KC-135's opposite wing, move into position, and take his fuel. With initial offload complete, we would cycle through again for our last 1,000 pounds. At least, that was the plan.

Like a high-speed elevator, my flight rose to the tanker's altitude and a bit behind him. Responding to my small rudder flutter, Parker slid out to a position just off the tanker's right wing. The first aircraft to refuel, I opened the refueling door. The night was now pitch black and there was no horizon. The lights of the tanker created my horizon.

I followed the "director lights" on the tanker's belly and moved forward into the refuel position and stabilized. The Boomer extended his boom and stabbed my aircraft. It latched. I watched my fuel gauge increase and confirmed fuel transfer. After what seemed a long time, the Boomer called "Disconnect" and withdrew the nozzle and boom. I dropped a few feet and smoothly moved to the tanker's left wing.

From the back, Jay snidely remarked, "Not bad for a slick-wing pilot."

I disregarded Jay's remark and watched Parker drop down and slide left to the Pre-Contact Position. Parker started forward, but then, his F-4 began to oscillate.

I radioed, "Johnny, settle him down."

"Two" backed out and after a pause, Parker moved to a good Pre-contact position, then the Contact Position, and received his 3,000 pounds of fuel with just a few bobbles. He disconnected and smoothly slid out to the tanker's right wing.

Our turn for the second offload. Jay asked, "Can I take this refueling?"

"Yes, but if you muck it up, and we have to eject, I **will** kill you when we get on the ground."

"Awww. You don't mean that!"

"Oh, yeah! That way, only one of us meets the Accident Board, and I can blame it all on you."

Jay chuckled and said, "Okay, I have the aircraft."

He gently shook the control stick and slid down and right, directly to the Contact Position. The Boomer plugged the refueling receptacle, and we took 1,000 pounds of JP-4 with only one slight bobble. I watched the director lights on the tanker's belly; they consistently showed Jay was doing well. The Boomer disconnected the nozzle, and Jay slid our F-4 over to our position off the left wing of the KC-135.

"Okay, Guy In Front. You can close the refueling door now, and you have the aircraft," said Jay, and he shook the stick.

"Done! Not bad for a slick-wing WSO. I have the aircraft." I shook the stick. (One snide remark deserves another).

On Parker's second attempt, he held steady, made contact, and received his 1000 pounds of fuel.

Then the tanker said goodbye to us. "Duddy 5-1 flight has its off-load and is cleared direct to Malaga and 2,000 feet, descend VFR, the altimeter is 30.11."

I read back the clearance and got a "30.11" from Parker. Our flight turned toward Malaga and rapidly descended to two thousand feet, transitioning to the local altimeter at Angels 06. Northward across Spain, we flew as a tactical two-ship to the first two turn points on the planned low-level route.

Reaching the planned split point, I radioed "Duddy 5-2 is cleared off to fly your planned low-level and attack." Parker acknowledged, "5-2 is cleared off to the left. See you in the debrief." With that, 5-2 turned left, dove into the blackness of central Spain, and began their night low-level route, ending with a simulated attack on an oil refinery.

Duddy 5-2 would recover as a single ship to Madrid and a planned TACAN approach to a GCA or ILS final approach and landing. TACAN is a Tactical Air Navigation system providing the user with bearing and distance to the station. (in this case Torrejon Airbase, Channel 94).

Jay commented, "I'll bet our new Lieutenant will learn a lot about low-levels and realistic tactical deliveries from Johnny tonight."

"Roger that. Okay, confirm heading and time to our next turn point."

We flew our low-level and then simulated a special weapons delivery over a brightly lighted refinery. We heard 5-2 radio "Off target." Eighteen miles west of us, and we knew they would begin a climb to contact Madrid Center. We called "Off target." and slowed to get more separation from 5-2.

As we got closer, I contacted Madrid Center, and Madrid cleared us to Flight Level 2-1-0 and direct to the Torrejon TACAN holding fix thirty miles east of the base. The few clouds present obscured the coastal lights near Valencia and then Madrid—hazy and no horizon!

My first 180-degree turn in holding was technically okay, but I felt uneasy. When it came time to roll out of the second turn, I moved the stick to the right. The instruments showed that I correctly leveled the wings on the outbound leg. But my head told me that I just rolled left to an inverted position!

I checked the instruments. There was no descent or turn; the heading was steady; but I knew I was upside down. *I should be hanging by my seat straps!* The primary attitude indicator still showed the gray sky up and the black ground down. As far as it was concerned, we were just fine.

"Jay, how is your attitude indicator working?"

"Looks good to me."

I was developing a twitch on the flight controls as my inner ear and eyes battled for control of the aircraft. It was time to confess what was happening. "Jay, tell me we are not inverted because my head feels as if we are."

"No, we are upright and wings level, on the correct outbound heading. Do you want me to take the aircraft?" Jay had taken over for pilots several times in Southeast Asia.

"Not yet, let me work this out." I glanced outside, but the blackness below and above did nothing to turn my mind right-side up. I told myself: *"Just fly the instruments. You have flown in thick weather before; this is the same thing. Believe the Attitude Indicator!"*

Just then, Madrid Center broke the silence: "Duddy 5-1, you are cleared for the high TACAN runway 2-3 at Torrejon."

I acknowledged the clearance. "Duddy 5-1 is cleared for the approach to Torrejon."

"OK, Jay, follow me through on the approach."

"Roger that. Thirty-degree left bank and maintain Flight Level 2-1-0. Roll out! You are wings level. You are six miles from the fix, call departing."

Traveling at five miles per minute, it was a nervous and shaky sixty seconds or so before I reduced power, lowered the F-4's nose, and made the radio call.

"Madrid, Duddy 5-1 is departing the fix and out of 2-1-0."

A short reply: "Madrid Center copies, Duddy 51,"

"On heading, good descent, descent checklist complete," Jay called. *There was still NOTHING to be seen for a horizon!*

I quickly ran my descent checklist, without moving my head for fear of making the vertigo worse. I still felt a tremor on the control stick. I was beginning to sweat.

As we approached 10,000 feet, Madrid Center called, "Duddy 5-1, continue descent to Flight Level 0-6-0 and contact Torrejon Approach, local Channel Five."

"Duddy 5-1, down to 0-6-0 and Channel Five. Gracias."

Jay immediately changed the radio to Channel Five and said, "You're up."

Out of habit I turned my head and looked down to my left at the radio display to confirm that I was, in fact, on Channel Five. *Damn! Bad move! My head felt like I made a hard roll to the right. I made a sharp left-roll control input.* Before the F-4 could roll 10 degrees, I looked back at the attitude indicator and aggressively snapped the wings level.

"Are you OK, do you want me to take it?" Jay called. He was following me on the flight controls. A jerky left roll, then a jerky right roll, tested his patience. He was ready to take control of the plane. If we went in, Jay would be the second aviator to the crash scene, 39 inches behind me.

Before I could answer, our F-4 broke through the bottom cloud layer, and there, thirty miles in front of us were the lights of Madrid, forming a horizon. Instantly, my internal gyros turned upright, and the vertigo vanished. *What I felt now matched what I saw.*

"I'm OK now, Jay. I've got the airplane. This has been fun, but let's go home."

Steve graduated from The University of Texas and earned a Master's from Auburn University. He served his country for 26 years as a Paratrooper, Green Beret, then Fighter Pilot. He published *"From Both Sides Now"* and has contributed to the *"TAC Attack,"* British *"Fly Past,"* MWSA *"Snapshots,"* *"Daedalian Flyer,"* and *"Reaper's Lament."* Steve lives in Texas with Claire, his wonderful bride of over 54 years. They have two grown sons.

Author photo credit to Sam Roberts Photography, Bulverde, Texas

CHAPTER 15

LISTEN UP

BY PAUL D. GONZALES

It was mid-September of 1967 that I received orders for Vietnam. I was assigned to Fort McClellan, Anniston, Alabama, Noble Army Hospital. McClellan was my first duty station after all my military training was complete as a Radiologic Technologist and a combat medic. I traveled with another private, "Fred." Fred was a tall lanky guy from Tennessee, kind of resembled a young Abraham Lincoln.

Fred and I reported to the hospital and met a Staff Sergeant "Jack" at the Personnel Office. Sergeant "Jack" took Fred and I to a long corridor and asked that we walk fifty feet make an about face and return to him.

Fred and I followed Sergeant Jack's order and executed the walk. When we returned, Sergeant Jack stated, "Gonzales, come with me. Fred, report back to the first sergeant."

I was assigned to the hospital staff, and Fred was assigned to an outlying clinic.

I thought to myself, here it is 1967, a Spanish-Italian kid from Pittsburgh, Pennsylvania, here in Alabama. "Just my luck!" Needless to say, I never left post alone for fear of meeting up with the KKK. My buddies from Michigan had regular names, so they would take me to their homes off post.

One Saturday we were at Chuck's house, a trailer where he and his wife Joan lived. We were drinking beer and listening to music, minding our own business when we heard a loud knock at the door. When I opened the door, there stood an elderly woman with a gun. She said, "If you don't turn that music down, I'll drop you where you stand."

With eyes as big as a full moon, "I said, yes, ma'am and welcome to Alabama, 1967."

I did good at Noble! Made rank rapidly (Specialist E-5), became highly proficient as an x-ray technologist and a combat medic, and received letters of commendation. My military career was fulfilling and rewarding. The thought of becoming a "Lifer" was entertaining. I was a medic and damn proud of it.

After several months at Noble, we were on first name basis, so I asked Sergeant "Jack" what was the walk down the corridor all about? Jack said, "I can tell by the way a man walks if he means business and has determination. I saw that in you Gonzales, so I wanted you on my team." Needless to say, my self-confidence soared.

Upon receiving orders for Vietnam, I was thrust into the reality of what military service was all about. Everyone in radiology was saddened about my leaving.

I felt indispensable, so I reported to the personnel sergeant and told him how valuable I was to the hospital and the radiology department.

The personnel sergeant sat there and listened intently without emotion and grunted once or twice and leaned back in that government gray office chair and exhaled.

After completing my speech on how valuable I was, the personnel sergeant rose, went to the water cooler, pulled a cup from the dispenser, and filled it with water. I felt my explanation of how valuable I was made him dry throated.

With the 5oz cup full of water the sergeant returned to where I was seated at his desk and placed that 5oz cup of water in front of me. I thought, how polite of him.

He spoke softly but firmly. "Gonzales, put your finger in the cup of water."

I was confused. "What for, Sarge?" I responded.

"Just do it, Gonzales."

I followed his order and inserted my index finger into the cup of water.

"Do you see a hole in the water?"

Again, I was puzzled. "Of course not, Sarge. "

"Then that's how much you're needed here. I will have you replaced in a week. NO SOLDIER IS INDISPENSABLE. DO YOU UNDERSTAND ME...GONZALES?"

Immediately, I understood his point and quickly recalibrated my military response and simply stated "Yes, Sergeant." I stood up purposeful and straight in military fashion and asked to be dismissed.

The personnel sergeant said, "Before I dismiss you, you have two weeks to gather your gear and sign off this post, followed by a thirty day leave before you report to Fort Lewis, Washington for deployment to Cam Rahn Bay, South Vietnam, UNDERSTOOD?"

"Understood, Sergeant."

Then in a clear but firm military voice, he said, "Dismissed, Specialist."

In that brief encounter with the personnel sergeant, I was transformed from a naive 20-year-old to a well-disciplined soldier. It left no doubt in my military mind that I was dispensable and replaceable. I was to follow orders without question.

Throughout my military enlistment and beyond, I accepted the reality of military: discipline and obedience.

Paul is a native-born son from Pittsburgh, Pennsylvania. He served in the U.S. Army from 1966-1969. He served in Vietnam in 1967-1968 as a combat medic and x-ray technologist. Paul served during the Tet Offensive of 1968. Paul pursued a career in radiology after his service in the Army and retired from University of New Mexico Hospitals after 22 years of service.

CHAPTER 16

THANKFUL: FINAL REFLECTIONS AT MY USAF RETIREMENT

BY RAYMUND M. TEMBREULL

"This is all very surreal. I've provided comments at so many retirement ceremonies as the presiding officer, but this is a first as the honoree and the last time as an active duty officer. To think of all the times, I whined about having to wear my blues, and now that this is probably the last, I think I am going to miss it.

"Strangely, I'm finding it hard to say goodbye to an organization I didn't even originally plan to give a career to. It's so dysfunctional. Also, a disclaimer: I've always thrown a lot of energy, passion, and emotion into what I do, so I have no doubt that will bleed through here today. Apologies upfront and viewer discretion is advised.

"When I thought about what I was going to say at my 'last military engagement,' I thought about what American citizens say to women and men in uniform all the time: 'Thank you for your service.' A simple enough statement; It acknowledges gratitude for our service and sacrifices of our families. Many of these citizens go above and beyond as anonymous donors, and I have been the benefactor of my share of free breakfasts and lunches or free drinks at the airport. Our country's collective sense of appreciation drove me to contemplate all those things I find myself thankful for at the conclusion of more than a quarter century of military service.

"It ended up being a pretty powerful list, and I'm going to share it with you ... but it needs to be kept close hold among those in attendance here today. I'm hoping that I don't expose some sort of military conspiracy, because upon review, I found I have a lot to be thankful for and much of it due to my military service. What I am thankful for ...

"I am thankful for my tribe ... for being a Defender. I love being a Defender. Internally, we can be as conflicted as there are differing perspectives about who we are and what we bring to the fight. We are a blue-collar career field in a white-collar service. To outsiders, we are 'a breed apart and make no sense' (to borrow a Last of the Mohicans quote). Popular myth has it that we eat our young, and in some respects, that is true. For young officers, we are literally thrown to the wolves. We are given charge of upwards of a hundred airmen often before we even attend tech training. We are 'issued' to a Master Sergeant to teach us our tradecraft, and our long-term success as officers is largely determined by the quality of our first Flight Chief in the formative first assignment. I was fortunate to have Lewis Stewart Jr. as my first Flight Chief. He hailed from the great state of North Carolina—'God's country' as he called it. If career progression is any indication of the quality of a leader, then Stew was it. He carved out the time to study for promotion every single day and made Chief Master Sergeant in 14.5 years. I also wanted to give a shoutout to Command Master Sergeant Jimmy D. Allen, the unit chief for my Operations Officer tour at Kirtland Air Force Base. We had a rough start but grew pretty tight...that Strategic Air Command-trained killer taught me graduate-level nuclear security survival. Chief Allen had his own way, and it was not a gentle one. But the skillset he gave me served me well during my command tours. Chief Allen beat a lot of things, including the Khobar Towers bombing, but he couldn't beat cancer. God rest his soul. I am very thankful for great enlisted leaders like Stew and Jimmy D. to show me the ropes, and what right looks like early on. They definitely started me off on the right foot and pointed me in the right direction.

"I am thankful for not just being a defender but also being an officer. Upon review, I will seem to contradict myself, because I also believe that the one thing that separates the American military from all the others … the one thing that makes us the most effective and lethal force on the planet is not our tech or our weapon systems … it's our Nonc Commissioned Officers, or NCO corps. And frankly, there's not much an officer can do that a seasoned NCO can't do better. The one thing the NCO corps needs their officers to do well is '**command**.' I have always felt it is a commissioned officer's obligation to command. I'm thankful for a career field that has given me multiple opportunities to serve in this important capacity. I hope I carried that crucible with honor and served my airmen and the mission well. I don't think anything else I will ever do professionally will be as meaningful to me as being a commander.

"I am thankful for the many places on this great planet my family and I have lived and experienced, because I was an airman. Five of seven continents … I've seen the absolute best this world has to offer, met the most wonderful people, and seen some of the most incredible sights and creatures. While I've had my share of bad days, I also know that I have been spared the world's worst and the horrors that many others have experienced for so long. At the end of the day, all we can really hope for is that we make a difference. For me, that was serving something greater than myself, and I am grateful for everything the Air Force has done for me and my family and all the opportunities I've been given. My life is forever enriched as a result.

"And when I think about the places, I really don't see things, buildings or grand vistas … I see the people. The people made these experiences really special. The people I've worked long shifts with in remote locations, broke bread with, laughed with, and sometimes cried with.

The people who believed in me when I didn't believe in myself. The people who followed me when I talked 'crazy' and asked the impossible, and they delivered every time. The people who saw the potential in me to serve in the next grade and assume greater responsibilities in the next challenging assignment. I'm thankful for the people … even the problem airmen and troubled dependents of assignments past. The Air Force is a family; you don't get to choose your family, but even at the end of the worst days, you're still together. Can't put a price on that …

"I am thankful for being able to work with *heroes* everyday—airmen, NCOs, senior NCO's, officers, civilians, even contractors—who put the mission and the well-being of their fellow airmen and those they serve with above their own every day. I've had the privilege of bearing witness to their heroism for over 25 years; whether they're pushing a police transition team in the chaotic urban sprawl of Baghdad; sweeping an indirect fire impact area at Balad, or making sure the last up-armored HMMWV chocked full of airmen makes it safely to their 'Remain Overnight' location during a North Dakota blizzard … or the supervisors who care enough to check in on the family of their deployed troops during their preciously scarce off-duty time. And even a brother from another mother who takes in my son as his own so he can finish his senior year of high school in the same place he started—*how do you ever repay something like that?* You can't … it's just what families do for each other.

And speaking of military families, I would be remiss in not mentioning them among my heroes too. I find the term "dependents" somewhat offensive, because we can't do what we do without their support. The Air Force recruits airmen but retains families. You certainly can't make a career out of the military unless your family is "all in" too. Their sacrifice and service is just as real as the service member's. We all bind together as a greater Air Force family to help each other through all the service-related absences and deployments. The key spouse who sits through the night with a first-term airmen's wife, comforting her as she struggles in a troubled marriage ... they're my hero too!

"I am thankful for everyone who took the time out of their busy schedules to attend this ceremony today ... many of whom have traveled great distances. Thank you for honoring my family and me.

"I am thankful for the small, yet mighty team who used their 'wonder twin' powers to form this memorable Air Force retirement ceremony to honor my family. Admittedly, I challenged everyone with an Air Force ceremony off the beaten path in Navy country ... asking for this to be piecemealed together over many months and from several locations. Bomb and Jonesy, thanks for everything you did on the front end. Routing my retirement medal and the magnificent shadowbox are among their handiwork. Chaplain Liu—awesome invocation. Thank you to my bride, Andrea who had to play protocol officer, script writer, program maker, event coordinator, and anthem singer. Thank you for putting this all together. Thanks to Chief Range for narrating this event and the team of Langley Defenders you brought with you to serve as proffer, honor guard, and escorts. You've constructed a magnificent 'air bridge' into Navy country today. Many thanks to the Shifting Sands staff. Thank you to everyone who has given their time to make this a memorable event for my family.

"I am thankful for today's presiding officer. Mr. Fedrigo is someone who has dedicated his life to serving his country: a great military career as a Defender and a chief master sergeant. He's served Defender nation in the highest possible civilian role and continues to serve the Air Force as a respected senior leader. You honor me, sir ... thank you.

"I wanted to thank both the Captain and Mrs. Kaseote ... Pop and Mom ... for their military service and the support they've given to Andrea and me throughout the years.

"I am thankful for my family and my blue-collar upbringing. 'Real talk' folks: my parents raised me right! Country-raised with a fair amount of manual labor required in the daily running's and upkeep of the Michigan homestead. Despite Dad working full time in the skilled trades and Mom being a nurse, they provided a rock-solid home environment where they instilled good values, inspired our creative side, ensured our academic success, and hardwired a strong work ethic into our DNA, along with the unassailable energy and passion to finish every job us kids have ever set out to do. A military career is a marathon and not a sprint. Upon reflection, I found that I had to draw on all of these attributes, along with firmly embedded empathy toward the plight of the working man, to be successful during the course of my career. Most importantly, I attribute my internal hardiness to grind it out and push through the hard times ... to never give up ... to my upbringing. Let's just call it that 'Tembreull grit.' I needed to draw those reserves quite a few times. I couldn't have done it without you guys! Beyond that, it's the travel to obscure locations for my promotion ceremonies. Don't think you missed one. The events' grandeur certainly didn't match the distances you traveled or justify the cost, but you were proud parents, and it was meaningful to me. Thank you for setting me on that path where I could do my small part for so many who have accomplished so much. I love you guys.

"If my life is a library, my mom and dad are the book ends, and my brothers, Vince and Tom, filled out the volumes on the top shelf as we learned to play, how to hunt, how to socialize ... all the things growing mammals need to be successful in the human race. I couldn't ask for better human beings to crew with. They are the type of guys that will give you the shirt off their back ... one need only ask. They are also great examples as husbands and fathers, with both managing to marry up in the process and raise remarkably normal children. I'm proud to call you brothers. Thank you to you and your families taking the time to make this trip. Andrea and I really appreciate it.

"I am thankful for my son, Alexander, who doesn't like being put in the spotlight, so I am going to do just that. I'm grateful for the precious

(Courtesy of author)

time we have spent together in the way a father who doesn't take it for granted only could. Despite all the moves, new places, life changes, and uncertainties, you have come through it like a champ. Thank you for putting your faith in Andrea and me even when you knew we were not offering the easy path. I am hoping retirement from the Air Force will add an element of stability that will allow us to link up for some more of those father and son trips I enjoyed so much. You were the best snowboard buddy and dive buddy a guy could ask for, and I was absolutely addicted to watching you play lacrosse. Thank you for all of those good times and great memories. My one wish for you is for you to find your passion and be passionate about it! I'm saying goodbye to one of mine today. Love you, son.

"And now for the hard part (which is why I saved it for last). I am oh so thankful for my wife and partner, Andrea. The Air Force and I have asked very much of her over my career. Seven assignments with at least ten different household moves over fourteen years. Of those seven assignments, four have been command tours, including our first assignment together in Guam. She went from dive instructor and her dog, Nikko, to officer's wife, commander's wife, and later (just add water … shake well) instant mom. You didn't have time to grow into any of these roles, and sure there were bumps in the road, but the proof you did a great job is right there next to you—our son. The proof of it is right here (point to rank). The proof of it is all around you. Thank you for always supporting me and believing in me (and I know you believe in me, because no matter how often or how bad I screw things up, you still keep setting ridiculously high standards for me). And thank you for always keeping me grounded and maintaining the right perspective: it's always been about the airmen. You've been their biggest fan, and their biggest cheerleader. You've experienced joy in their successes and sadness in their stumbles and struggles. You've breathed life into my military service and my approach to leading. Thank you for sticking with me … then, now and forever. You are proof that not all stars belong to the heavens. Sometimes they come from the sea. I love you.

"I look around this room and see the quality individuals and great Americans that have served and are serving our great Air Force now, and It does my heart good. I am thankful for having been a part of such a worldclass organization. Because it is a strong institution, it will continue to attract the very best of America's sons and daughters. It is the reason I am able to walk away from it today.

"So, to those who say, 'Thank you for your service,' I say, 'you're worth it!' I have served a great nation made up of great people. And on the down-low, it was a pretty good deal for me too.

"Again, thank you everyone for coming."

Retired Colonel Raymund Tembreull is a devoted husband and proud father, a 26-year combat veteran of the U.S. Air Force, and an accomplished career professional in law enforcement, physical security, and antiterrorism. In his professional endeavors and personal travels as a naturalist and scuba diver, he has experienced humankind's extremes as well as the planet's most amazing places and creatures. He employs his creativity at every opportunity, and his passions are eco-themed literature and art.

CHAPTER 17

SPIRITUAL UNREP

BY PAT ALDERMAN

UNREP is underway replenishment when a ship pulls up next to tanker while underway (cruising along at a rapid pace while at sea). Spiritual UNREP is a way for sailors to get a mental or spiritual nourishment while at sea. Based upon a 30 October 1965 letter to family and friend, during USS *Midway's* first Vietnam cruise, written by Captain James O'Brien. "To date, the *Midway* had conducted 114 UNREPS, Spiritual UNREP religious services on the flight deck."

Somewhere
in the middle of nowhere
out on the deep blue sea
a Chaplain held a service
for sailors just like me
No arching cathedral overhead
just nature's bright blue sky
a memory that will stay with me
until the day I die.

No limits shall be placed on me
I can see to the horizon
a feeling that I hope to retain
'til I am old and wizened.

> Thus far this cruise, MIDWAY has conducted more than 140 underway re-plenishments (UNREPS) with other Seventh Fleet ships to sustain our extended operations at sea. Perhaps the most rewarding UNREP came on Sunday, August 8 when we held a "Spiritual UNREP" on the flight deck. On this day Catholic Mass and Protestant Worship Services replaced the blast of jets and roar of engines. Attendance at these services set a new high for the cruise. In a different kind of UNREP, the handling of tons of ordnance is an exacting and back-bending job. Praise came on August 21, when crewmembers of MIDWAY and the ammunition ship PYRO teamed up to shatter a Seventh Fleet proficiency record by transferring an average of 176 tons of ammunition per hour.

(Image courtesy of author)

After thirty years as military librarian serving in both base and academic libraries, Pat Alderman retired from federal service, moved to San Diego with her retired naval officer husband, and began volunteering on the USS *Midway* (CV-41) Carrier Museum in the library. They now live in Charlottesville, VA where she continues to volunteer remotely for the *Midway* by copying deck logs from the National Archives and cataloging.

CHAPTER 18

VENTAIL

BY MICHAEL LUND

When anyone asked him about his nose shield, Charles would chuckle, "I'm one step ahead of the pandemic requirement to wear masks." The nose shield was a little flap that attached to his glasses' bridge between the two lenses and hung down over his nose. "A trend setter I am," claimed Charles, "a man who sees the future."

Donna, his wife, moaned. While it was true that others his age often seemed obsessed with the past, he did have ways of looking forward. And at least he generally did such bragging outside of her presence.

She was self-conscious of being in the Covid "vulnerable" category (seniors), and tended to stay inside except to chat on the phone or at an appropriate distance across fences with her neighbors. It was not until their daughter's wedding that she understood why Charles wore the Ventail even after his face was healed.

He'd started wearing it after he underwent a procedure to remove potentially cancerous growths from his face. It involved rubbing a cream on twice daily that attacked suspicious blemishes. It would basically burn off unhealthy cells, leaving pale, pink skin underneath.

"You look like you've been baked in the oven," Donna had claimed after the first week of treatment.

"I do," admitted Charles.

He was looking in the bathroom mirror, turning his head to the left and then to the right. "Just another week. But I *am* going to be a sight even afterwards." Still, he was no more self-conscious about his looks than usual.

When he first went out for his daily walk with a splotchy red face, though, he worried about sunburn on the tender areas. Given Charles' Scandinavian genes and a childhood spent outdoors on a farm, his dermatologist had recommended wearing long sleeves and pants in general and, in particular, a hat with a broad brim. But the regular trips out to talk with other veterans was important. He discovered a solution online, the Ventail.

The name was inspired by the piece of a medieval helmet that protected the lower part of a knight's face. The modern device, specifically designed to block ultraviolet light, was like a Groucho Marx vaudeville prop, but without the bushy eyebrows and mustache.

When Donna teased him about becoming a clown, he said the ventail was grander than a costume piece. "This," he pointed, "is inspired by the noble Don Quixote, suiting up to do good deeds in a world of scoundrels. He was a man who not only looked back to the golden age of chivalry, but anticipated the evils of the Industrial world, those giant windmills."

"Is that the kind of soldier you were? A rebel who went out on his own to right wrongs?" Donna teased him. They had met ten years after he got out of the Army.

All Charles had told her about that phase of his life was that he was one of the military's millions of clerks. She knew and that he'd done a tour in Vietnam and spent the last year of service at Dover Air Force Base in Delaware. But he never talked about it.

He sniffed: "I kept meticulous records of equipment orders and shipments."

As far as she knew, he had kept up with just one friend from the service, Stephen Block of Charleston, South Carolina. They would talk by phone once or twice a year and occasionally send clippings about politics. But the past they shared was a mystery to Donna.

She did wonder if his part in a war, even if not as a combat soldier, prepared him to endure other hardships. He insisted the challenges they personally endured in the pandemic were minimal. They had a secure retirement income; their house was paid for, their lifestyle simple.

When their daughter Linda became engaged, though, Donna lamented to friends, "He wants to go shopping for a wedding dress with us, that thing on his face. For all I know he'll wear it to the ceremony."

At The New Olde Bridal Shoppe, Charles advised Linda, "The sleeves should lie comfortably about the elbows."

"The elbows?" wondered both Donna and Linda.

Their son Franklin, in his rebellious teenage stage, wore mismatched socks, let his pants hang in the back to reveal brightly colored underwear, and sported a scraggly, patchy beard. He intended to wear his signature outfit rather than a tuxedo for his sister's wedding.

Charles moaned. "It's like someone threw clothes at you from a bin of random seconds. Apparel should be connected, if not unified."

Franklin countered that Charles should add a codpiece to his attire, a flap of material that attaches to the front of a man's trousers and covers the genital area. "If you like Don Quixote, why not imitate him further? In Cervantes' time, codpieces were quite popular. They've even made a resurgence with metal rock bands." He raised an eyebrow, "With, that is, the leather subculture."

Charles raised an eyebrow himself. "Well, no danger of sunburn there."

He had other quirks regarding his appearance. When visits to barber shops were off limits, he asked Donna to give him a haircut but insisted on directing her and measuring the results.

"Where is a three-way mirror when you need one?" Charles complained, turning the hand mirror which showed him from the back in the full-length hallway mirror. "Maybe I need to use modern technology."

Thus began a series of goofy experiments involving his iPhone on a tripod, his laptop on the kitchen counter (where there were adjustable lights set in the ceiling), and a security camera mounted in the corner.

"I hope you're only examining your head when you do that," Donna cautioned. "If you want... um... other inaccessible parts looked at, please ask me."

"No need," he countered. "The full-length mirror is satisfactory. And it's the view from the front that matters."

"At least disconnect from the Internet when you study yourself," Donna begged. "I don't want some government agency to think you're working on an identity theft program--or planning something kinky."

She was allowed to make only a few snips at a time with the scissors. "I don't want curls showing off my neck. And the top should be even, thin as it is."

"You could shave your head," Donna said.

He responded with a non-sequitur: "Do you have any AA batteries?"

"I do. Why?"

"My electric toothbrush plug-in system seems to be failing, but it can run on batteries."

Calling it an "electric toothbrush" was an understatement: it was a computerized apparatus with multiple functions. Charles programmed it to brush for sixty seconds at a time and moved it at the end of each cycle to a different area of his teeth (top left outer, top left inner, etc.). It also included a sonic brush and a water jet. The twice daily operation took over ten minutes.

He bought replacement bristles every two months—well, he did before the pandemic. Fortunately, he had planned ahead and stock-piled several replacement packages.

"I told you," he said to Donna when she saw him pull a new set from the drawer. "I see the future."

When she confessed this to their daughter, Linda wondered if his antics could be signs of approaching dementia. But Donna knew he was too sharp in other respects: working the daily Sudoku with blaz-ing speed and analyzing the daily stock reports in a way that impressed their financial advisors.

"Why does he brush his teeth that way?" Franklin asked Donna. "Is he panicking about losing his teeth?"

"I don't think so," his mother admitted. "He doesn't even get cavi-ties. And he has the dentist whiten his teeth every year."

Donna felt Charles had never done anything with his appearance to impress people professionally. A data analyst for a national insur-ance company, he seldom interacted with customers or other staff directly. Reports came to him from regional offices, and he processed the information to produce bar graphs, spread sheets, pie charts, and narrative summaries.

Donna didn't mind missing office parties because of his reserved manner, but she might have enjoyed some of the cruises awarded to top performing personnel (he earned them every year). Still, he was exceptionally kind, attentive, and sensitive to others' needs.

"Charles seems like a machine from the outside," she admitted to her own social circle, other (mostly retired) health care workers. "Data in, data out." She would hesitate. "But, you know, he's not like that at home."

Franklin and Linda agreed. "He's spontaneous, funny, uninhibited even."

"Your dad?" their friends would respond.

"I'll give you an example. When we were young, he taught us how to spit shine shoes—the Army way."

They would ask, of course, if they had worn that kind of shoe.

"Oh, no. He loaned us each a pair. But the process turned into a game of fast draw. We'd see who could tie their shoes the fastest. And he seemed so serious—'ready, set, tie'! His face was set, his eyes boring into the tips of the laces. But when we were done, he'd laugh and laugh."

"So, your dad had multiple pairs of leather shoes. Did he have one for each day of the week?"

Franklin would deny it, though to himself he admitted it might be true. He did know he could see his face in his father's polished shoes.

Because of continued Covid restrictions, the wedding party had to be small, and the ceremony was held outside with only immediate family and a few close friends. His old Army friend Stephen Block sent a package that arrived the day before the ceremony. Franklin put it in the dining room with other gifts.

Donna opened the carefully wrapped box without thinking and found a forty-page booklet tucked under a slow cooker. The cover featured a picture of two soldiers standing in front of a sign, "U.S. Army Mortuary, Da Nang, Vietnam." One of the men could have been Charles; the other was African American.

A note was sticking out between the pages. "I know you never liked talking about those days, my friend. Your Midwestern powers of repression are admirable. But, a Southerner, bottling it up for so long, I had to come to terms with what we were asked to do by writing about it. There was a program at the VA."

The two men, she read in the preface, were "mortuary affairs specialists," the equivalent of civilian undertakers. They identified corpses, preserved personal effects, and prepared the remains for burial.

Donna fell back into a chair. Charles had done that? And his efforts to display the body's beauty were how he coped?

She knew that in recent wars, reporters were not allowed even to film coffins being unloaded from military aircraft. It was all to be kept from public view in order to avoid adding to anti-war sentiment. These two men were confronted with the reality of death every day and did what they could to ease the grief of families. Were they ever thanked?

Stephen also wrote: "You were by far the best at what we did— attention to detail, respect for the dead, compassion for the families. I know it took a toll on all of us. Be safe, be well."

She decided to hide the book until after the wedding, where she cried as Charles gave their daughter away, Franklin scowled, and Linda pledged, "I do."

Michael Lund is the author of novels inspired by Route 66, America's Mother Road, and currently is working on a novel series set in a small coastal Carolina village. Lund also directs Home and Abroad, a free writing program for military, veterans, and family, in rural central Virginia. He was a U.S. Army correspondent at Fort Campbell, Kentucky (1969-70) and in Vietnam (1970-71).

CHAPTER 19

I HATE PROFESSOR DOBBS

BY HARRY MAYER

Naval Air Station Lakehurst, New Jersey, 1979

Despite bright mercury lights, it seemed dark in the old blimp hangar. Flapping pigeon wings echoed in the large empty space. When I looked up, several birds were joining the flock that had made their home on the rickety catwalk. The wooden catwalk was a death trap, 120 feet above the floor, with missing boards and a broken handrail. But this didn't bother the pigeons; they happily built their nests near the massive hanger doors. Just below the catwalk white bird droppings splattered the old walls too high above the floor for maintenance workers to clean. This gigantic building once housed the Navy's Zeppelins that patrolled the East Coast hunting submarines. It felt like I had stepped into an old black and white war movie, and I half expected to see Gregory Peck in a leather flight jacket briefing the pilots in the Ready Room.

Dressed in a green flight suit, I entered the Ready Room and hung my aircrew survival vest over the back of a chair. A row of squadron cups hung on a pegboard near the coffee mess. I didn't rate a squadron cup yet because I was only an aircrew candidate. Once I earned my wings, I could hang my cup on the wall with the others. I poured a cup of stale black coffee into a Styrofoam cup and took a seat to wait for the Pre-Mission Brief to begin. I arrived an hour early since this was my first flight as an enlisted aircrew candidate, and I was filled with apprehension. With only two drill weekends under my belt, I still had a lot to learn. One of the things I liked about the Naval Reserve was that I could earn extra money during the week to work on my qualifications, the perfect part-time job for a college student.

With time to kill before the briefing I pulled out a sealed envelope from my helmet bag. I had been struggling in Professor Dobbs' English Literature Class. I didn't particularly care for Professor Dobbs. She struck me as a pedantic woman overly obsessed with social justice. With her blond hair tied in a tight bun and wearing a frumpy sweater. her eyes would light up when she talked about the anti-war movement of the 1960s. She could turn any discussion into an obscure lecture on the evils of the American defense industrial complex.

While I usually read her assignments, I rarely participated in class discussions. I found the course material boring. Our most recent assignment had been a 5-page essay on *The Role of Madness in Hamlet: Insanity or Strategic Ploy*. I read every analysis of the Danish prince I could get my hands on. I was sure I nailed the assignment this time.

I took a deep breath and tore open the envelope to review my grade. In bright red ink across the top of the page was the letter "D" with a frowning smiley face. Then I read the comments, "Mr. Mayer, while this is certainly a well written paper, I have a sneaking suspicion that someone else wrote it." My face turned redder than the ink on the page. I muttered, "That bitch!" Petty Officer Perdue, who was sitting in the row of seats ahead of me, turned and said, "Did you say something?" I sighed, "It's nothing." I stuffed the paper back in the envelope.

Petty Officer Perdue, a first class petty officer with over 1,000 hours in Sea King helicopters was my NATOPS instructor. After the mission brief, I followed Perdue to the flight line. A blast of hot jet exhaust from returning Helicopter 552 hit us in the face. Its burning

JP-4 smelled strong and pungent. Helicopter 552 engines screamed on deck while its large rotor blades beat the air. A plane captain, wearing a cranial head protector and goggles, signaled the returning flight crew to shut down as we headed to our aircraft. We conducted a walk around pre-flight inspection of our bird. I read the checklist, and Perdue showed me what to inspect. Once onboard I took the right seat near the window, Purdue took the inside one. He showed me how to plug into the Internal Communication System (ICS), and once I did, I could hear the pilots talking in my helmet. Perdue said, "Tonight's going to be simple. We're doing some touch and goes, and if we have time, we'll do some practice with the hoist. Are you up for riding the wire?" I nodded.

I listened to the pilots recite the actions from the takeoff checklist. Then a voice from the tower said, "5-5-5 you're clear for takeoff." The helicopter aircraft commander brought the engines up to speed and pulled back on the collective. We took off at twilight. I watched out the window as we rose into the early evening sky. The red and white checkered water tower and the enormous Lakehurst hangers grew small as we climbed above the Naval Air Station. I marveled at the beauty of the Pine Barrens at dusk, the sky's vibrant red, blue, and orange colors slowly fading to black as the sky darkened. Soon the red lights came on inside the cabin as we flew into the night.

Perdue called to the pilots, "Pilot-Sonar. Request permission to unstrap to conduct post takeoff inspection."

"Permission granted."

Perdue unstrapped from the seat and meticulously inspected the interior of the aircraft. He inspected the control cables, hydraulic lines, and the integrity of the air frame. Then he strapped back in his seat. "Pilot-Sonar. Post takeoff check complete." For the next three hours the pilots practiced their take offs and landings. Then a voice came over the ICS, "Sonar-Pilot. You guys want to get some practice with the rescue hoist?"

"Sonar-Aye."

Perdue signaled me to follow him to the back of the helicopter. He handed me a gunner's belt and showed me where to fasten it to the air frame. Once we were properly hooked up, he slid open a large door at the back of the helicopter. After we were in a stable hover, he hooked me up to the hoist and then pushed me out of the door. I dangled out of the helicopter suspended on a tiny wire while he lowered me to the pad. Once I hit the ground I unhooked and waited for the chopper to return. As our flight was coming to an end, we practiced auto rotations, an emergency procedure for engine loss. Once at altitude the pilots disengaged the rotor head, and we dropped from the sky. It felt like the first large dip on the giant roller coaster at Great Adventure Amusement Park. Although this was my first flight, I was hooked. I knew I was going to love this job and couldn't wait to come back.

I returned to college eagerly looking forward to the next drill weekend. I had a paper due on Friday on *The Tale of Two Cities*. I sat at the desk in my room and stared at the Underwood typewriter. I searched for words to fill the blank page, while Bruce Springsteen's "Thunder Road" played on the radio. Just as I started to clack and clunk on the manual typewriter, my train of thought was interrupted by the WJRZ news announcer. "Good afternoon, ladies, and gentlemen, this is George Jessup, we're breaking into our regular programming with this developing story. A naval reserve helicopter from Lakehurst Naval Air Station, on a routine training flight crashed in a farmer's field near Bordentown, New Jersey. Navy officials reported earlier today there were no survivors. While the cause of the crash is unknown, the Navy promises a thorough investigation. Our thoughts are with the families of the flight crew. Stay with us for continuing coverage of this story. We now return to our regularly scheduled programming." Even though I hadn't been in the squadron long enough to know the members of the flight crew, I had a sinking feeling in the pit of my stomach. I lost all interest in my essay. Up to this point, danger seemed abstract, like it would happen to someone else. For the first time, I realized I could get killed in the reserves, even in peacetime.

The next day, I waited with the rest of my class for Professor Dobbs. Our English Literature Class was supposed to begin at 10:00 a.m., but she still had not arrived. By 10:15 students started to leave. A long-haired kid in an Army fatigue jacket said, "Man, I lucked out, I'm glad she didn't show up today. I was partying last night, and I never completed the paper. What about you?"

I said, "Hell yeah, I was working on it until midnight." The next day I received a call from the HS-75 Duty Officer who told me there would be a memorial service at the Cathedral of the Air in Lakehurst on Saturday morning.

Just beyond Red's Tailor shop on South Chapel Road stood the Cathedral of the Air. A beautiful Norman-Gothic chapel nestled among the pines. Sailors gathered in dress uniforms outside the church. A blanket of auburn pine needles covered the ground. A cold November breeze blew brown oak leaves across the parking lot. We waited in respectful silence for the memorial service to begin. Black limousines carrying grief-stricken family members started to arrive. I was surprised when I saw Professor Dobbs get out of the second limo with her teenage son. She was dressed in a black dress with a black veil. She took her seat in widows' row at the front of the chapel. Soft, colorful light filtered into the sanctuary through the chapel's magnificent stained glass windows. The atmosphere was melancholy, yet peaceful as the organist played "Nearer my God to Thee." I took a seat in the pew near the back.

Captain Reynolds, the Commanding Officer, walked to the podium. He said, "Ladies and gentlemen, officers, and men of Helicopter Anti-Submarine Squadron Seven Five. We are here to celebrate the lives of Commander Douglas Dobbs, Lieutenant Commander George Chadwick, and Petty Officer First Class Amelio Santangelo. Over the course of my career, I have attended too many memorial services. These brave young men, struck down in their prime only wanted to serve their country. They were America's best, and I felt proud to personally know all three. These were men of character who cheerfully answered their country's call to duty when asked and served with honor in Southeast Asia. We mourn their loss with great sadness and pray we are worthy of their sacrifice."

Following the Skipper's remarks, he invited family members, colleagues, and friends to speak. Professor Dobbs walked to the front of the chapel. Despite her bravest attempt to look stoic, she appeared fragile. She wiped away tears with a handkerchief. There was a long pause. She started to cry when she began. She said, "I'm sorry." She paused again. Then she spoke, "Today, we are here to celebrate the lives of three wonderful men, George, Amelio, and of course my Doug." Her voice quivered.

She continued, "Doug was not only the great love of my life, but he was also my best friend. Memories. So many memories. I remember when I was in graduate school, and he was a midshipman at the Naval Academy. He looked so handsome in his uniform. We would sit for hours on Sunday morning at Naval Bagels on Taylor Ave. He always knew the right thing to say; it was such a happy time. He would make me laugh when he spoke of some of the mischief he got into at the Academy. More than anything, he wanted to fly. His eyes would light up when he talked about flight school. He couldn't wait to earn his pilot wings.

"When my oldest brother John was drafted, and later killed in Vietnam, he comforted me. Doug was always there when I needed him. When I gave birth to Jason, he sat by my side in the delivery room for a day and a half holding my hand. He was my rock. I felt safe in his arms.

"My heart sank the day he broke the news to me that HC-1 was deploying to Vietnam. He told me not to worry; he would be safe on the *Ranger*. I was scared for him. Dougie had the heart of a poet and wrote me some of the most beautiful letters, I'll treasure those letters forever.

"When he returned from deployment he had changed. He had soured on the Navy and just wanted out. Once his initial service obligation was up, he submitted his Intent to Resign. I was ecstatic when he received approval. Once the war was over, he became restless. He told me how much he missed flying. The worst fight we ever had was when he told me he was going to affiliate with the Naval Reserve. He still wanted to fly, and I was furious. But that was Doug. Once his mind was made up, there was no changing it."

With tears streaming down her cheeks she concluded, "Doug, you broke my heart, it's so hard to say goodbye. I miss you, there is a piece of me that's gone forever. My dear, you've left this world, but you will never leave my heart." Then she broke down and cried. Her son hugged her and helped her to her seat.

There was a long silence in the chapel. The silence of empathy and sorrow. The service concluded with the Chaplin's benediction and the playing of "Eternal Father" with its slow and dignified melody, but it brought little comfort.

I returned to school on Monday and resumed classes. I never mentioned I was at the memorial service; there was nothing I could say that would change anything. Professor Dobbs never spoke of the accident either. She continued teaching and remained outspoken on social issues. Even though she continued to be a fierce critic of the armed forces, I found I no longer hated her. I continued to be bored with her assignments, and as you might expect, my English Literature grades never improved. But I no longer saw her as pedantic, only as a woman suffering a great loss. I admired her. She had been robbed of what she cherished most in this world, and nothing would bring her husband back, her life forever changed. I respected her strength and courage. She just pressed on.

HS-75's aircraft were grounded pending the results of a safety investigation. Some aviators wanted nothing to do with flying after the crash. We waited for several months for permission to resume flight operations. Three enlisted aircrewman signed their Page 13's to document that they no longer volunteered to fly. Some of the pilots even resigned their commissions. I couldn't blame them; most had joined the reserves to earn extra money. Now that the war was over, they felt safe returning to the Navy. The thought of dying from a part-time job was more than they had bargained for. When the safety investigation was completed, the cause of the crash turned out to be a cotter pin. A simple cotter pin had been incorrectly installed and vibrated off in flight causing the helicopter to fly out of control into the ground from 500 feet. They didn't stand a chance. Five months later our sister reserve squadron on the West Coast, HS-85, also had a fatal crash. One of the main rotor blades came off in flight, and all three souls on board were lost.

It's been forty-five years since the Bordentown crash. Though the events of that day have been lost to time, the accident still haunts me. I often think about how a simple cotter pin changed the trajectory of so many lives. In my case it was the realization of my own mortality. I could have been on that aircraft just as easily as the ill-fated crew of Helicopter 555. It's the luck of the draw; life's unpredictable that way. It doesn't matter how careful you are, when your number is up, it's up. And there is nothing you can do about it.

In late autumn I often sit on the deck behind my house and watch the Canadian geese. From my deck I have a nice view of Oliphant Lake. It astonishes me how those big clumsy birds can drift weightless on the air currents then lightly touch down in the water barely making a splash. Sometimes when I'm out there I'll hear the roar of military helicopters overhead. I'll look up and watch them returning to Joint Base McGuire-Dix-Lakehurst. The sound of those gigantic helicopters take me back in time. It evokes memories of a time in my life when jet engines were screaming above my head, and I could feel the vibrations from those giant rotor blades slapping the sky. In those quiet moments of reflection, I think back to a time when I rode in those magnificent mechanical monsters and remember how thrilled I was to fly above the clouds just for the chance to soar like an angel.

(Images courtesy of author)

Harry's career spanned from the United States Marine Corps to the Navy, serving as a Deep Sea Diver and Bomb Disposal Technician. After twenty-three years in the military, he coordinated medical teams during catastrophic disasters. Harry's writings, inspired by his experiences, explore leadership, crisis management, and the human spirit, reflecting his profound service commitment.

CHAPTER 20

HUMOR IN SERVICE

BY KATIE STRAIN

"He who laughs, lasts." -Mary Pettibone Poole

The sun beats down on my body. It's far too hot to be standing in the direct sun in the middle of summer, but here I am.

"Can we get a picture with you Ms. Prairie?"

Laughter erupts from the male soldiers standing in front of me. I adjust the giant skirt that has a built-in hoop and frown, a smile being stifled.

"You know you want to laugh!" One of them says with an arched brow.

I do. I want to bust out laughing, but I hold it in.

"Fine," I say, adjusting my bonnet.

We take a few pictures, and I hand them the "Apple Days" brochure.

"I can't believe you got tasked to do this! The chapel had these costumes? You look just like Laura Wilder from that one show..."

"*Little House on the Prairie!*"

"Yeah, that one!"

Finally, I can no longer hold it in.

"Shut up!" I laugh.

"Why did first sergeant lend you to do this tasking anyways?" They howl.

"Hey, I pull duty for the chapels on Sundays, I have to work on the weekends too!"

"Haha! I'm glad to be an engineer then!"

We laugh at my plight, and I offer them the snacks set up inside. I was the *only* unlucky female 56M (chaplain assistant) in the post that could fit into a prairie gown from the local historical society. Fort Riley was putting on its annual "Apple Days" that is open to the public, and they required the main post historical chapel to be open so history could be showcased. Someone had the brilliant idea of making a few of us walk around in costumes.

What got me through such a humiliating experience as a young specialist? Humor. Humor would be a huge part of my time in the Army. And it helped me get through so many situations to come, ones that would be fraught with darkness and hardship. My job, or military occupational specialty (MOS) was a chaplain assistant. I would see the worst of situations. Suicides, suicidal ideations, assaults, interpersonal issues, relationship issues, anything you can imagine, I would hear in counseling sessions and in visits to the troops.

But humor would be there. As it always is.

"Did we get them?"

I pant, struggling to hold the water balloons in my hands.

"Yeah! Give me another one! Hurry!"

My boss reaches down; his eyes fixed on the view over the Hesco barrier we are standing behind.

Someone shouts, "Who is throwing water?"

I start to laugh and once I do, I can't stop.

"Let's go."

I follow suit, moving as swiftly as I can, with over 40 water balloons balanced in my untucked physical training (PT) shirt. A few fall, splashing onto the dirt and soaking my shoes. But I don't care. I am *happy*. One of the rare moments during the deployment I have felt this way. And the deployment has been hard. We have seen too many improvised explosive devices (IEDs). There have been too many fire-fights and countless hours on the routes being in danger.

"It's the chaplain and Sergeant Strain!"

"The UMT?!"

I hit one of the guys squarely on his head, causing everyone to bust out laughing.

"She got you!"

They figured us out. We pelt our guys sitting around the firepit and laugh until tears come out of our eyes. They laugh too, a good respite from the day of hunting IEDs. And we had also gone out that morning with another platoon. I needed this just as much as them.

"How did you get these filled up?" A sergeant first class asks us, astonished at the volume of balloons we managed to chuck at them.

"We went in the empty women's shower trailer, propped the door open and used the sink!" I reply.

"You guys are dedicated, Sh*t, we appreciate you!"

That memory is forever etched in my mind as one of the ways humor relieved some stress from serving in combat. The unit I found myself in was definitely the most unique of my career, and it impacted the rest of my time in. If not for humor, I don't know how I would have handled those fifteen months. And it seemed to be universal.

Fast forward to the unit after, once I came back to Fort Riley and the deployment was over.

"There is a bat."

"A *bat*?"

"Yes, Sergeant. A bat, the literal kind. It's flying around in the sanctuary."

"What in the…"

I found myself in a garrison position, managing the largest chapel on the base. And currently, Catholic mass was going on, about halfway through the service, when my soldier jogs into my office.

"Yes. It's in the middle of the sanctuary, buzzing people's heads!" He whispers as we approach the doors to the sanctuary.

Swallowing a lump in my throat, I push open one of the swinging doors slowly. Praying that I don't get bit and contract rabies, I walk in. Heads turn, and a bunch of worried eyes meet mine. A bat is indeed buzzing heads as my soldier said. He was correct. Sometimes junior enlisted like to pull your tail, so I had doubts if it was real or not.

"What do we do??" He whispers again, from behind.

Motioning for him to follow me, we skirt the side of the nearest isle and watch the bat. The Catholic chaplain continues to give his address, unphased by the creature flapping around during his service. He is a no-nonsense Filipino man, and I know he knows it's there, but has full confidence in me to get rid of it. More confidence than *I* currently have if I am honest.

Thinking quickly, I direct my soldier. "Go get the long communion baskets."

My soldier jogs to the back and brings me the baskets, with long poles attached. Thankfully, we have these. It should work. I hope anyways.

"Ok, you go to one side of the aisle, and I'll get in the middle of the pews…we work together to catch it."

The sermon is still going as we walk slowly toward the area where the bat is flying. I hear a few young kids giggle and say how weird we look.

"There!" I whisper.

Hesitating, I raise my basket in the air.

"Get it Sergeant!" My soldier loudly whispers.

Standing in the middle of the pews, people all around me, I touch the bat for a split second and whimper. Soft laughter rings out, and the chaplain keeps talking. Thankfully by the grace of God I stifle curse words and inch past knees as I keep my eyes on the bat. My soldier struggles to catch the bat by himself as it flies in the middle of the main aisle. I am pretty sure at this point it sees us and considers it a challenge to stay airborne.

Only it keeps flying, erratically at this point. We can't catch it easily. Frustration begins to set in as I move in between people, interrupting the service with our movements. I can feel anxiety begin to take over so I stop, walk to the middle isle and whisper to my soldier.

"I'm gonna call animal control. Stay here. Make sure it doesn't bite anyone."

I quietly leave the sanctuary and walk back to my office.

"Animal control."

"Hi, this is Staff Sergeant Strain from Morris Hill Chapel. There is a bat in the sanctuary."

"A bat?"

"Yes, a bat. Can you send someone to come and get it please? We have Catholic mass going on at the moment and…"

Laughter rings out through the phone and I hear more than one person laughing too.

"Sorry, just picturing that. We will send someone."

If it wasn't up to me to fix the situation, I would have been laughing as well. But alas, this was all on me as the chapel manager and noncommissioned officer in charge (NCOIC). Just as I hang up, my soldier appears, with a box.

"I caught it!"

"What?"

"It's in the box!" He says, cheerily.

Scratching sounds and squeaking come from the box.

"I had to step on a few people's feet though, Sergeant. I didn't mean to, but I had to, to get the bat trapped in the offering baskets."

I watch the box move on my desk, and a laugh escapes my lips. Then he laughs. Then we *both* laugh. Deep belly aching laughs.

"And the chaplain is still preaching! He didn't react at all!"

The omission makes it funnier, and we laugh until tears come out of our eyes.

Fast forward again to a small village in Germany.

"Sir, I can't make it in on time. I will have to crawl out the kitchen window to unlock the chapel!"

I was assigned to a large brigade on the small base set in the city of Bamberg. But as a chaplain assistant noncommissioned officer, I still had to rotate on Sunday duty at the chapel to help out garrison. And this morning, I was about to be late getting to the chapel for Catholic mass to start.

"Katie, calm down, what are you saying? Why can't you get there on time?"

The Catholic priest liked to call me by my first name when no one was around. He was an unconventional and chill guy.

"There is a giant *spider*! It's hanging over the door! I can't get out or it will fall on me!"

I was and still am deathly afraid of spiders. No matter the size. But this one, this one was a huge wolf spider, and it was the literal size of my hand!

"I am on my way! Stay where you are!" The chaplain laughs.

I knew he would help, and I was tremendously grateful. His car motors up the single path driveway and stops behind mine. I stand frozen in the hallway, unable to move or take my eyes off the arachnid beast. The windowpane on the front door was muted so only shadows could be seen from the other side. But I can make out the silhouette of the chaplain, and he held something long in his hands.

"Katie, it's me! I'm coming in!"

Slowly, the front door opens and there stands my rescuer. He has on his Kevlar, safety glasses, long leather gloves and is holding a long lacrosse stick with both hands. If you have never seen one, they have a basket attached to the end. Suddenly, I let out a giggle. He was the funniest sight I had seen in a long time. He chuckles too, as he angles himself under the doorway and slowly maneuvers the giant spider into the lacrosse stick basket.

"You weren't kidding! This is a big boy!"

"Careful!" I say, watching him ease the stick outside so he could fling it into the field beyond my tiny basement apartment.

"EEEEKKK!" He shouts, sending the spider airborne.

I can hold the laughter in no longer, we both laugh so hard that tears come from our eyes. Once we calm down, he turns to me.

"You know I am telling everyone at mass about this right?"

Nodding, I accept my fate. It is worth the incoming laughs at my expense to have my house rid of the spider.

It wouldn't be the last time laughter took over a very stressful situation. Humor would go on to help me through countless ranges, being stuck on duty during holidays, getting through more situations that involved serious counseling and then some. Humor is one of the best

The photo above is of the author, SGT Katie Strain. Iraq, 2006 (Courtesy of the author).

kept secrets to surviving military life, to staying sane while sent away to far off places that you find yourself in. Military life is hard, and it only gets harder when you add deployments in the mix.

I am grateful that I served the country I love and grateful for all the good people I had the privilege to serve with. I see my service as one that will outlive myself and that lives on with the junior soldiers I had then and those that they had underneath of them. And I hope I was able to give them the gift of humor. The magic key to enduring whatever the military will throw at you. I will forever carry humor with me as a United States Army veteran. And one truth rings out.

Everyone needs to laugh.

The photo above is of the author, SSG Katie Strain. Bamberg, Germany, 2012. (Courtesy of the author)

Katie Strain is a signed thriller author with Provoco Publishing, under the pen name of KE Jennings. Katie also is a contributing author in the all-female veteran anthology: *The Haunted Zone*, out with Tundra Swan Press and a contributing author for the Armed Services Arts Partnership's 2024 anthology: *Partnerships, The Untold Stories*. Katie is an Army combat veteran and enjoys spending time with her son, skateboarding, and hiking, as well as reading at home.

CHAPTER 21

TIME IS FLEETING

BY MICHAEL TYSON

:00

I feel it there, stuck below the xiphoid process, snuggled in my costal margin, like an annoying pebble in my boot at the start of an eight-mile ruck march, unreachable, inaccessible, and inconvenient. Perfect. Maximum coverage. I got here in time. Now the wait.

:01

This must be how the princess felt. How many mattresses did she feel the pea through? Been many years since I read that story. Mom was so mad when I put those frozen peas under my mattress and forgot about them. But Dad thought it was funny. How the hell could the princess feel one pea under so many mattresses? Sounds like bullshit to me.

:02

Damn this thing, poking me in the gut. Reminds me of the pain after eating Aunt Donna's oyster stuffing last Thanksgiving. Painful knot in the stomach, hours of lying down praying for death, finally solved by prolific shitting and cold sweats. No worries about that this time around. This knot is external, soon to affect my internals. This frigging thing. Digging into me like that fat bastard Johnson's elbow on the flight over. Why do I always end up in the middle seat? Just because I'm new? Nine hours! Damn that *rank has its privileges* BS. Couldn't get a wink of sleep. I'll get plenty of sleep now.

:03

So uncomfortable for such a small thing. Like a baseball. That big? Maybe the size of a racquetball ball? They are called pineapples, after all. Feels bigger than it really is. How's my vest up and above it? Gotta fully cover it. Why's it feel so huge? And my vest so small? Like that snug shit-brown vest under that cheesy disco-era suit Simon made me wear as best man. At least he's got a life. And I'll be an uncle soon. Sorry to miss that. Dammit. Where will this thing feel better? Nowhere of course. I'll move my arms, close off the openings, smother it, give it no out. As if that can really do anything. But maybe it's a dud? Probably not, my luck. Time to close my eyes and pray. Oh God, the dirt. Forgot about the dirt. What a nasty smell. OK, gonna calm down now. Slow my breathing.

:04

How much time has gone by? Jesus, why is this taking so long. Is there time for a Hail Mary? I've been lying here forever. Is it always like this at the end? Why is my life not flashing before my eyes? Is there a deeper meaning? Perhaps my life is full. Complete. Done. Maybe this one really is a dud? Russian grenades are known to...

:04.2

Approximately 350 cast iron fragments, traveling at 24,000 feet per second from the detonation of the 110-gram charge of TNT, are cut short in their deadly mission by the concrete floor, body armor produced by the lowest-bid contractor, and the body of Private First Class Henry "D-Bag" Debouchette, originally from Pierre Part, Louisiana, and most recently of Fort Hood, Texas. A handful of shrapnel makes its way through the late private's forearms and right elbow, embedding themselves harmlessly in the northwest and southeast walls of the house he and his squad just recently breached.

:05

Lieutenant Wilson feels the blast before he hears it, like waves hitting the shore at Surfside Beach. He rides the concrete floor as it buckles, his body rising on the crest and coming down with a thud in the trough. A dull popping sound reaches his ears. He had expected a louder bang, but now knows why it never came.

:07

Then, the screams of his men. He knows what they'll find and wants to stop them from seeing what they are about to see but knows he can't. He doesn't have the right. They have to see what has been sacrificed for them. He hears one of his soldiers vomiting.

:10

He gets up and joins his men. The platoon's war-hardy staff sergeant is standing over D-Bag crying, unashamed. Private Sorenson is kicking the mud-brick wall, screaming. The rest of the platoon are kneeling around their former friend. Wilson looks at the body. It's really bad. Poor bastard. Nineteen's too damn young. The beginning of the citation comes to mind easily. "For conspicuous gallantry and intrepidity above and beyond the call of duty..."

Michael Tyson is retired Air Force, married to an Air Force vet, with whom he's had four children, three of whom are veterans. The smart one works in insurance. Michael was last published in the FAO Journal of International Affairs describing his experience as an Air Force NCO in Berlin the night the Wall fell.

CHAPTER 22

THE PINE WOODS OUTSIDE BASTOGNE, DECEMBER 1944

BY DENNIS MAULSBY

A great sigh. The Sergeant's breath
condenses into ice on his mustache.
Give him another morphine.
The Medic blows into fists
formed by clinched bare fingers.
He's had two already; a third will kill him.
The Lieutenant moans, whites of his eyes roll up.
Both his stumps twitch.
We got the tourniquets on too late.
Give him another; let him go out
in pleasant dreams far from this hell.
Needle stabs, images form. *I—I—I....*

I walk among rain-soaked lilac bushes.
My shoulders brush the flowerheads.
Water shakes off to drench my uniform.
My wife waits, arms half-raised. Her wet blouse

translucent, perfumed, arms ascendant.
My mind free as a child. Fear, anger,
cold—a vague memory. From the farmhouse,
the scent of chicken and noodles finds me.

A collie's cold nose and warm tongue
kiss my palm. A happy whine.
I am home. I am home. I am home.

Dennis Maulsby's poetry and short stories have appeared in numerous journals, including *Star*Line, The North American Review, Haiku Journal, The Hawai'i Pacific Review, The Briarcliff Review,* and National Public Radio's *Themes & Variations.* His award-winning books include three collections of poetry, two volumes of short stories, and a novel. A number of his poems have been individually published in online or print journals in Sweden, Switzerland, China, India, the United Kingdom, and Russia. Learn more at his website: www.dennismaulsby.com.

CHAPTER 23

AL ANBAR AIR

BY INGO KAUFMAN

During Operation Iraqi Freedom, the CH-47 Chinook helicopter was the coalition's shuttle service in and out of Iraq's western province of Al Anbar.

Although I stepped aboard this workhorse only three times, each trip framed my seven-month deployment into distinct phases: a step into the unknown, a near-fatal jaunt to the decadent capital, and a welcome return back to the hellhole I somehow came to miss.

Early in 2004, I left Kansas with about ten other soldiers to join our brigade, already five months in-country. After a three-day trek to Kuwait, we caught a C-130 to Baghdad. The steep takeoff pinned us to the seats, and the combat landing left my diaphragm straining to keep my stomach down. As someone explained how the abrupt angles minimized the window of vulnerability to certain threats, I pondered how helicopters coped with these dangers.

While most of the group's destination was Ramadi, two of us were headed for a different base on the way. We waited for a flight several days in a mini tent village pitched in a remote corner of Baghdad International Airport. I chatted with the captain and enlisted men in my group, and they taught me how to play Spades. When my beginner's luck ran out, I read, wrote letters, and emailed friends and loved ones.

One email etched in my memory was from a lieutenant friend back in Kansas who wrote our unit had lost another platoon leader in combat. I had a hunch where I was headed.

We finally left late one evening. A truck drove us to a dark parcel on the airfield where two Chinooks waited. Their rotors churned the air and dust into a miniature sandstorm. Since arriving, I'd learned they usually flew in pairs and at night. I stepped out, grabbed my gear, crossed the tarmac, climbed into the front helicopter, and sat back against the starboard wall. As we ascended, I turned to peer out the little window. Faint outlines of fields, palm groves, and clusters of buildings dimly lit by sparse lighting came into view. It was my first trip ever in a helicopter. The ride was smooth and uneventful—just the way I like it.

Well after midnight, we descended into pitch blackness and landed on a desolate airstrip in Al Taqaddum. Only two of us got out: a private headed to a different company on the same base, and me. We hoisted our packs and duffel bags, descended the back ramp, and stepped onto the tarmac.

Weighed down by my gear, I scuttled through the downdraft of wind and dust as the metal birds took back to the skies to destinations further west.

It was an overcast, cool, and breezy February night. I was the officer-in-charge. My companion looked to me for guidance, but I had no idea where to go in the fathomless darkness that enveloped us. I probed the blackness for the faintest glimmer of light in any direction. Nothing shone save a few faint stars from between the clouds.

Prior to leaving Baghdad, lacking any contact information and leaving nothing to chance, I'd asked about transportation on arrival.

"Your battalion's expecting you, sir. They should send someone." The tone conveyed classic bureaucratic dismissal and groundless assurance. Further inquiry was pointless.

We were alone in the middle of the desert, and no one was coming. My vision adjusted enough to make out a rocky ridge lining one horizon and barren landscape on every other side. My companion asked, "What do we do, sir?"

"Get off this tarmac and look for a road. But be ready to sleep here."

Fifteen minutes later, I was about to deploy my sleeping bag. We found a dusty route, but it went two ways and was nearly indistinguishable from the surrounding terrain. It was wiser to wait for first light than risk getting lost.

Suddenly, a pair of headlights appeared in the distance and grew bigger as a vehicle approached. Someone *was* coming after all! We advanced and flashed our flashlights, determined to be noticed. A Humvee turned and parked just off the tarmac. The driver stepped out.

"What are you guys doing out here?" a female voice asked. Her name was Staff Sergeant Bluebird. She wasn't there for us, but we had our ride. "I'll take you back, but I wasn't expecting anyone, sir. The Chinooks that brought you took contact on the other side of that ridge—rockets. I'm on my rounds. Did you hear or see anything?"

We hadn't. Fortunately, the rockets all missed. We climbed into the Humvee, rode back to camp, and spent the night in considerable comfort on soft couches in the Morale, Welfare and Recreation area. The next morning, my companion and I caught a small convoy to our destination where we parted ways to join our respective units.

As the months passed, mission blended into mission, and the greenness behind my ears faded. My experienced soldiers taught me a lot, and I learned quickly. I kept cool under pressure and earned my place leading a cohesive team that operated as one.

One hot day in late spring, we set up a traffic checkpoint on the highway. We were good at it and learned to use terrain to our advantage. The viaduct five hundred meters to the west made us invisible to approaching eastbound traffic. Two of our Humvees hid behind the rising ground on each side of the overpass, and by the time oncoming drivers realized we were ahead waiting, our companions were already behind their flanks. It was a classic mousetrap inspired by interstate speed traps back home, and I was proud of it.

We randomly selected cars to pull over onto a small turnoff that held about five vehicles, ordered the driver out, and searched car after car for "contraband," inquiring in broken Arabic: "Where are you coming from? Where are you going? 'Bonneet' (hood). 'Sanduk' (trunk/boot)." These routine missions rarely turned up anything of vital interest.

But one did: a trunk full of weapons and ammunition, including a Dragunov sniper rifle with armor-piercing rounds. The driver was a Syrian with a deeply lined face and a red and white kaffiyeh on his head. He looked over fifty but was probably barely forty. He was calm and indifferent, resigned to his plight.

We detained him and impounded his vehicle. The next detainee convoy delivered him to Ramadi, and from Ramadi they transferred him to Baghdad.

This was the last we heard of the affair until mid-summer, when the company commander produced a subpoena summoning me and another witness to the American Embassy in Baghdad to testify against a weapons smuggler.

That August, almost six months after my first ride, I was back at the same landing strip boarding another CH-47 Chinook. It was just after midnight. A half-moon rose over the eastern horizon.

I wasn't alone this time either. One of my soldiers accompanied me: Pfc. Saul Gomez, a hard-charging Puerto Rican.

The Chinook lifted, the cabin vibrating with the whir of the massive rotors bearing me and a handful of passengers into the sky. We were headed to Baghdad with a stopover in Fallujah. As always, the Chinooks flew as a pair. Ours brought up the rear.

We were buckled into our seats in the front half of the aircraft's portside, our backs to the little Plexiglas portholes. I twisted and craned my neck to look out. The half-moon shone like an unwelcome beacon, casting enough illumination to betray secrets better off kept hidden.

After barely fifteen minutes airborne, we landed in Fallujah to pick up more passengers. We took off again and flew low. I watched tensely as the ground passed by beneath us.

On the eastern outskirts of Fallujah, a series of muzzle bursts confirmed my apprehension and briefly revealed a group of sinister figures lurking below in a field. Tracers tore through the darkness as the gang of hostiles fired their AK-47s at our aircraft.

"Gomez, we have contact downstairs!" I hollered over the din.

My stalwart companion glanced over his shoulder. Sparks hailed down outside past the windows as some of the rounds ricocheted off the rotors. He crossed himself in fast-motion and kissed his crucifix as a smoke trail from a rocket-propelled grenade rose and arced just over the top of the chopper. A moment later, a second one with a corrected trajectory headed straight for us. The pilot took evasive action, and the chopper pitched and rolled like a trawler in a tempest. A tremendous blast beneath us sent a concussion through the hull.

"This is it," I thought, commending my soul to my Maker as I waited for smoke to fill the cabin and braced for the inevitable fall. But we kept flying, and there was barely any smoke.

Relief gave way to anger, then frustration. I was alive but enraged, imprisoned in my seat with a window I couldn't open. My weapon was useless.

As we flew on, I took a deep breath, intent on recognizing how fortunate we were. Providence, professional pilots, and a marvel of engineering kept us aloft. The Chinook was a beast that could take a beating.

We arrived in Baghdad's Green Zone late that night. Passengers from both choppers mixed as we walked off the tarmac. "Holy shit, did you see that other chopper get lit up?" one of them said, mistaking me for someone else in the dark. "I thought they were fucked. I'm glad we were in the lead one."

"I wasn't. I was in the one you thought was fucked."

"They always aim for the second one," someone else quipped.

Gomez and I hadn't slept in twenty hours, but we weren't tired. We'd just left the tarmac when the sirens screamed, and loudspeakers advised everyone to report to the nearest shelter. It was a mortar attack. In the confusion, we asked directions to the embassy.

At the next checkpoint, tensions were high. Marine guards asked if we were there for the "excitement" a few minutes earlier. I shook my head as we pushed through to register. Once our bags were in a tent, we walked behind the palace that was now the American Embassy to check out its famed swimming pool. The mortar attack alarms sounded again. Again, we had no idea where these shelters were and there was no one to follow.

"Gomez, this will probably be over before we find a shelter, so I'm inclined to lay low right here."

"Me too, sir."

We sat poolside as a few mortars cascaded down. The intensity was short-lived, subsiding into intermittent lobs several minutes apart. The closest one landed about five hundred meters away, near the banks of Tigris. Neither of us was generally complacent, but we relished the moment, specks in the gigantic Green Zone, basking in our invincibility because we both knew Death wasn't coming that night.

We struggled to find the little courtroom the next morning. Lost in a remote and quiet corridor, we surprised two flirting officers who begrudgingly ceased their ass-grabbing long enough to provide directions.

In a fast hearing, we validated the inventory list from his trunk and confirmed that the suspect, an unconcerned-looking, middle-aged Syrian, was the same individual we'd detained.

With thirty-six hours to burn before our flight "home," we swam in the pool and looked up a local contact who graciously showed us around the Green Zone. This was our sole opportunity for tourism in Iraq. We visited the Monument of the Unknown Soldier and the crossed Swords of Qadisiyah, a monument formed by arched swords brandished by fists rising from a base of broken enemy helmets and spanning two ends of a large square.

Although we enjoyed a well-earned break and decent food, I couldn't wait to leave the Green Zone. In two days, I'd seen enough. Service members rubbed elbows with contractors who earned more money in a month than my whole crew put together. Civilians and contractors could buy and consume alcohol, but for us, the military's General Order Number 1 forbade us from enjoying a single cold one. In one large hall, we passed the entrance to a restricted area where the senior military officers and civilian leaders had their offices—guarded by private security.

The place felt depraved and artificial. Moreover, it didn't feel right lounging poolside while my men were sweltering out in the desert subsisting on lousy food and risking their lives. I think Gomez felt the same. There were two worlds in Iraq, and the sparkling but degenerate Green Zone imparted an air of wholesomeness to the dusty, rural province we'd left behind. The country mice were ready to head home. Al Anbar wasn't Kansas, but at that moment there was no place like it.

Our return flight was the kind I like: boring and uneventful. I don't even recall if I heeded the advice to take the lead chopper. Baghdad's suburbs stretched out below us like a beaten vagrant lying in the dust. As we flew past plumes of smoke rising from little red spots smoldering below, I was happy to leave the capital behind and return to Al Anbar.

We landed back at the familiar airstrip a few hours before dawn in much the same circumstances as my first flight into the sector: I was standing there with a private, someone was supposed to come pick us up, and once again, the tarmac was empty.

But this time was different. I knew exactly what was going on. The soldier next to me was like family. Our platoon—brothers we could count on—would come for us.

We also had night vision. Gomez and I found a spot near the road, made ourselves comfortable, and waited.

As countless stars twinkled across the sky, a waning crescent thumbnail moon rose over the eastern horizon. A meteor blazed overhead and faded. A moment later, there was another. More followed. It was the annual Perseids meteor shower, and this was my first time ever witnessing it. For an hour—perhaps the best one I spent over there—I lost count as we watched shooting stars hail down from the heavens.

Just as the eastern horizon shimmered with the promise of a new day, we spotted the headlights of our platoon's Humvees in the distance. They approached and pulled up alongside us.

My platoon sergeant, Sergeant First Class Raymond Navarrette, rolled down his window. "Sir, Gomez, we're heading straight into Khaldiyah for an all-day mission. We'll brief you on the way."

As I stowed my rucksack in the trunk and climbed into my familiar, worn, and uncomfortable seat, I smiled inside, my heart pulsing with life and my soul with gratitude. My feet were back on the ground where they belonged. I was back where I belonged.

And while I would always hold CH-47 Chinooks and their crews in the highest regard, I hoped never to set foot in one again.

Ingo Kaufman served in the U.S. Army from 2003-2006, including a stint as platoon leader in Iraq with 1-34 Armor, 1st Infantry Division. Since then, the Michigan native has worked in various operational roles for two Fortune 500 companies on two continents. Ingo enjoys nature, reading, and spending time with his wife and two children. He is nearly finished with his first book, a collection of interviews and personal accounts from his time in Iraq.

CHAPTER 24

A TRIBUTE TO A LEADER

BY WESLEY R. HILLSTROM

My dad, Wesley Rudolph Hillstrom, spent his childhood in Salo, a Finnish farming community in northern Michigan, near the town of Hancock. A remarkable lad, at the age of thirteen, he placed first in a county-wide scholastic contest and won a trip to meet the governor of the state of Michigan. He enjoyed listening to classical music and learned to play the accordion and the piano. From his early years, he was placed in roles requiring sacrifice and service.

His adolescent years, during the Great Depression, were difficult. His father, a carpenter by trade, and his older brother, the oldest of nine children, often traveled out of state to find work to provide for the family, which left him as the oldest sibling, essentially to run the farm. He was thus unable to continue his education beyond eighth grade. All through his life, he remembered the school bell ringing, heralding the first day of a new semester, as he walked with tears in his eyes behind a horse-drawn plow.

He later found work digging ditches in Texas. There, the penalty for leaning on a shovel or pick longer that it took to light a cigarette was immediate dismissal; a replacement was then chosen from a line of men waiting for a chance to work. On one occasion, he had a very painful toothache and asked the foreman for permission to see a dentist. The foreman appreciated his work performance and gave him one hour to see a dentist and return to his job. Dad was in such a hurry to return to his highly lucrative ten-hour, seven-day, $28.00 per week job, that the dentist pulled the wrong tooth before pulling the right one. He returned to work within the hour. That evening, he also had a part-time job playing accordion in a tavern. Such experiences contributed to his development of mental and physical toughness and in the development of inner strength and a strong work ethic. Those qualities undoubtedly contributed to his success as an effective leader. From a young age, his favorite poem was *Invictus*, by William Ernest Henley, a poem that illustrates pure courage, ending with "I am the master of my fate: I am the captain of my soul." During his early life, that poem was his creed.

As the Depression years ended, he found steady, well-paying work as a gear tester at the Timken company, a manufacturing plant in Detroit, Michigan. In Detroit, he also met and married the strikingly beautiful Madeleine. Life was going very well, but all was to drastically change. Shortly following his wedding, he was drafted into the Army. As a skilled machinist, he could have been deferred from military

Early photograph of the Hillstrom family home, built by Dad's father, Joshua Hillstrom, circa 1926. (Copy of a photograph courtesy of author)

service, but he chose not to apply for a deferment. His attitude, like millions of men of his generation, was that he was ready to serve if his country needed him. As his parents came from Finland, all that he knew of American history was what he had learned in school; yet that was sufficient to instill in him a sense of patriotism.

He arrived for basic training with two distinct advantages: farm-developed strength, and a good knowledge of firearms, since hunting was a large part of his youth. He was assigned to an 81mm mortar platoon, in the 96th Infantry Division and adapted well to military training. He also took correspondence courses to augment his education. Once, he was admonished by the drill sergeant after making an error while marching. With quick wit, he stood at attention and exclaimed in a military manner, "Sergeant, the intricacies of these military manipulations are far too complex for my diminutive mind to comprehend!" The drill sergeant simply ordered him to fall back in line.

Mom and Dad in their wedding photograph, May 23, 1942. (Copy of a photograph courtesy of author)

Dad quickly rose to the rank of platoon sergeant. He recalled an early address by the incoming platoon leader, in which that lieutenant stated that he expected every soldier of a given rank in that platoon to know more than any soldier below that rank, and that he would make it his business to know more than Sergeant Hillstrom. Dad silently smiled; he thoroughly knew the military manual and assumed the lieutenant would have some difficulty in accomplishing that objective. To his surprise, Dad said that the lieutenant succeeded by devising a method of firing mortars in a battery configuration. His platoon leader was an excellent student at West Point and had proven himself, prior to and during combat, to be an excellent leader. On one occasion, later in combat, the platoon was caught in a banzai charge. Without hesitation, the lieutenant ordered my dad to put a mortar in action and ran toward enemy, ultimately calling for mortar fire practically on his own position, thus allowing his men to withdraw safely. His first thought was for the safety his men, an action that earned him a Bronze Star.

Dad later learned a valuable lesson, regarding his ability to be the "master" of his fate, as per the poem *Invictus*. During one combat situation, utter chaos prevailed. Ultimately, everything just seemed to come together to where order was somehow re-established without his direct intervention. He concluded that, with respect to any situation, he had relatively little control over circumstance and could only do his job to the best of his ability and let life unfold.

Early in combat, Dad demonstrated effective tactical initiative. Following a grueling march, the platoon leader ordered him to make camp near the edge of a wooded area. The lieutenant was required to leave the platoon to get orders for the next day and gave Dad the command of the platoon. After the lieutenant left, Dad began surveying the area with binoculars and began to feel very uneasy; he thought he saw a glint of light in the distant hills. He reasoned that if he were defending that area, he would already have the ranges of certain positions calculated; those areas could later be shelled with mortar or artillery fire. He expressed his concern with the platoon section sergeants, but their consensus was that the lieutenant knew what he was doing, and they should remain where they were.

Disobeying a direct order under combat conditions, without good reason could result in a court martial. If he were wrong, moving the men from where he had been ordered to make camp could put them at further risk. Still, Dad could not shake that dreaded feeling. Just before dark, he gave an order for everyone to move 200 yards into the woods and re-dig their positions. According to Dad, the air "was blue with cursing" from the men, but the order was obeyed because they had been so well-trained by their platoon leader. When nightfall came, the men were exhausted. The entire night, the area they had left was devastated by artillery fire. The next morning, from a safe distance, Dad showed the men the devastation. From that time on, none of his orders were ever questioned.

Dad's platoon leader thought very highly of him. On one occasion, the company headquarters received an intelligence report of an imminent Japanese attack. As the time and location of the attack were unknown, a directive was sent to the company that four of the best men from each platoon were to be stationed around the company perimeter to specifically draw the enemy fire and alert the company. The lieutenant read the directive and ordered Dad to select two of the best men in the platoon and report to him. Dad complied and then questioned the lieutenant as to why he had requested only two men. The lieutenant replied, "Well, if you and I aren't the other two best men in this platoon, then there's something wrong with it."

Dad had interesting combat experiences. Late one evening, he felt something, perhaps a snake or centipede, bite him in the right shoulder. He was unable to call for a medic, as orders had been given to shoot to kill anyone that moved or spoke, after nightfall; he had no choice other than to wait in silence until dawn. At first light he called for a medic and was taken to an aid station. After minor surgery, he was given one shot of penicillin every hour for thirty-six consecutive hours, to counteract the effects of the poison. Dad considered that episode to have been a blessing in disguise. His feet were highly infected with "jungle rot," as he couldn't take his boots off for several weeks, for if he did so, his feet would swell, and he wouldn't be able to again wear his boots. Fortunately for Dad, the penicillin not only counteracted the effects of the poison but also healed the jungle rot.

On another occasion, as Dad was fluent in Finnish, he was asked by another soldier of Finnish descent to help him write a letter to his mother. On the way to that soldier's outfit, he was met with machine gun fire. Dad intuitively knew his only chance for survival was to run in a zigzag fashion toward the gun, for the gun was on a hill. And if he could get close enough to that weapon, he could run underneath the line of fire. One of the sergeants in Dad's platoon, who was overlooking Dad's run from a nearby ridge later told him that as he was running,bullets were churning the ground between his legs. Dad said the bullets came so close to his head that they sounded like cracking whips. He successfully got under the line of fire, made his way to the soldier's outfit, and helped the soldier write the letter. Later, as Dad was returning to his outfit while running with another soldier, a bullet hit that soldier's rifle butt, as it crossed in front of my dad's side.

Dad was awarded the Bronze Star for digging at least four soldiers out of a cave during a continuing barrage of artillery fire. On another occasion, he was offered a Purple Heart but respectfully declined the award. He was struck by a spent bullet but felt there were too many soldiers who had paid too dear a price for that medal for him to accept it for such a minor wound. That was Dad.

To say the least, Dad was a responsible soldier. Nevertheless, soldiers have been known to celebrate, even on the battlefield. For my dad, May 23, 1945, was such an occasion, as that was the date of his third wedding anniversary, during a time when his outfit had been fighting for control of a certain hill on Okinawa. That evening, he and another of my dad's buddies, the company machine gun platoon sergeant, were celebrating by drinking a medical alcohol, lemon powder mixture. That next morning, he woke up in an ambulance, trying to find out what had happened the night before. In exasperation, a medical officer finally gave him the choice of either returning to the States to a medical pension or returning to the battlefield to satisfy his curiosity. Dad chose to return to the battlefield.

According to his drinking buddy, that evening, Dad had run to the top of the hill, waving a 45 caliber pistol and throwing handfuls of mud in the direction of the enemy, while yelling at them to come out of their holes and fight like men. His buddy, who was essentially as inebriated as Dad, ran after him and dragged him back down the hill. Miraculously, no one, neither American nor Japanese, fired a shot during that episode. Dad's company commander told him that, had he been sober, he would have recommended him for the Medal of Honor, and then said he should order him to pick up empty beer cans. Dad and I never discussed that incident at length but I'm reasonably certain he was probably thinking that he should be home with his beautiful wife instead of on some island halfway around the world. With that in mind, he probably thought, "Let's just get this blasted war over with and go home!

Dad told me that anyone who claimed to have been in combat and said that he wasn't scared was either lying or crazy. He related to me an incident wherein he found himself alone in the dense jungle, after catching up to one of the men who had run from battle and convincing him to return to the front line. Dad had a strange experience, wherein he was caught between two powerful inner suggestions. One was that if he returned to the front, he would be killed, and that he should therefore wait until the fighting had ceased, then say he got lost in the jungle; the other suggestion was that it was his duty to return to the front line. Dad said his feet felt like lead as he, with effort, had to physically place one foot in front of the other to return to the battle. He added that after he had overcome that fear; it never again returned with that level of intensity. Later that day, his platoon leader asked him if he would accept a battlefield commission to second lieutenant. Dad answered affirmatively.

Four days before dad's commission was to take effect, one of the sergeants in his platoon was killed while doing the job of an officer who was unable to perform his duties. Dad recalled the mother of that sergeant pleading with him at a train station before leaving for overseas saying, "Please take care of my boy." That really got to him. With justifiable anger, Dad went into the officers' club and strongly admonished that officer for not performing his duty that resulted in the sergeant's being killed. That action cost him his commission to second lieutenant. Dad's company commander told him that he was probably the best platoon sergeant in the company. But if he were to let him receive that commission, he could never give another order as a company commander. Dad was offered a choice to either remain with his outfit as a private or be transferred in grade to another one. Out of a strong sense of loyalty, his answer was that he had trained with those men, and he would remain with them. He was reduced in rank to a private, but shortly afterward, he accepted an offer of an assignment as recon sergeant and was given a sergeant first class rating. In retrospect, it was probably better that Dad didn't get that commission. My understanding is that the attrition rate for officers on Okinawa was quite severe. Had my father received that commission, I might never have known him.

Shortly after that incident, the platoon leader was wounded by machine gun fire. Dad, under a barrage of phosphorous shell fire, with the help of another soldier, made a makeshift stretcher and carried the lieutenant off the battlefield to an aid station. The last words the lieutenant said before passing out were, "You're doing a fine job, men, give em hell!"

Following Dad's return to civilian life, a fellow worker at Timken's told him that while he was "playing around out there" he paid for his house, while working long hours as part of the homeland defense effort. It took Dad a great deal of restraint not to clobber his clueless colleague; the man obviously did not fully understand the relationship of his privilege of earning good wages to the service and sacrifice of men like my dad and those in his outfit.

Dad pursued mechanical engineering studies on the GI Bill but couldn't seriously continue. He had PTSD and tried to deal with the effects of the war by drinking. Our family had some rather uncertain times. Dad ultimately quit drinking and helped many other individuals attain sobriety as well. My mom played a significant role in helping Dad to adjust to civilian life. She was a gentle petite lady, who had a reputation as a family caregiver on both my mom's and my dad's side of the family. As one of six siblings, she was no stranger to the hardships of the Great Depression. After the United States entered WWII, she and her younger sister worked as volunteers for the Red Cross, folding bandages. I remember my dad, with tears in his eyes, tenderly praising my mom who stood by his side through their nearly fifty years of marriage and ever remained the girl of his dreams. Dad passed away in 1992, and Mom followed in 2004. To my last day, I will be grateful to God for allowing me the privilege of being their son; they each epitomized service and sacrifice.

The 96th Infantry Division, or the "Deadeyes" as they were known, was a highly distinguished outfit, comprised of a unique brotherhood. For years following the end of World War II, members of my dad's platoon would meet at three-year intervals, generally coinciding with the Fourth of July. The reunions became occasions wherein the Deadeyes and their families would meet and enjoy quality time together at various locations within the United States. One year, the reunion was hosted at our family home in Detroit, Michigan. I can still recall the genuine camaraderie that the former "Deadeyes" enjoyed, forged in sacrifice and service. They all contributed to the service of the "Greatest Generation" in the defense of our nation, and I am proud that my dad was one of those unique men.

Wesley Hillstrom grew up in Detroit, Michigan. He lives in Mississippi as a retired geophysicist from the Naval Oceanographic Office. He has a B.S. in Geology and a M.S. in Marine Science. He volunteers at the Manning Family Children's Hospital and at the National WWII Museum, in New Orleans, Louisiana. He serves on committees for the 96th Infantry Division Association Heritage Fund and is a member of Toastmasters.

CHAPTER 25

DOG DAYS

BY BOB RITCHIE

The rain fell at an angle, steady but light, pelting the Quonset hut turned living quarters. Jake leaned back in an old aluminum chair under the small patio roof, his feet propped on the railing. An open book lay face down on his outstretched legs. In every direction, the Appalachian hills stretched like silent sentinels holding The Sanctuary in their protective arc. All along the northern edges of the pasture towering ferns and rhododendrons soaked up the weight of the storm, holding the pasture grass in check. Wildflowers bloomed in patches, poking through the damp earth like small bursts of defiance. To his left he could see the rain-soaked horses grazing in the southern pasture. They seemed indifferent to the soaking, tails swishing, heads to the ground. He knew they preferred the blanket of gray overhead and the wet soil beneath their hooves to the doldrums of their stalls as much as he appreciated the patio roof over his head. Here the rhythms of life moved in slow deliberate beats. It was neither soothing nor unsettling; it was just beginnings and endings. For a while, he could just exist without the weight of the world trying to drown him.

Jake's mind wandered back to his first weeks at The Sanctuary, nothing more than a broken man, a felon on the run from a life he couldn't bear anymore. He'd been offered this place in lieu of a jail sentence—a gamble by a veterans' treatment court to keep him from spiraling further into the wreckage of himself. He was skeptical of anything that sounded remotely like therapy. He hadn't believed in healing—not for someone like him. He couldn't shake the memory of nights spent on the streets, homeless, alone, angry, and desolate. The stink of old whiskey, stale cigarettes, and every unsolved thought he buried deep clung to him like a shadow of regret.

But then there was Sandy. A Vietnam Vet who had endured his own hell. Sandy's quiet presence shaped The Sanctuary into something more than a program—it was a second chance. Early on, Sandy had handed Jake a weathered copy of a wisdom book he simply referred to as the Tao, with nothing more than a gruff, "You'll figure it out." He hadn't pressed for answers, not even when Jake failed to answer his simple question: *"What was your best day?"* Instead, Sandy left the silence to do the work. And slowly, as the days passed—rain, shine, or fog—The Sanctuary became more than just a place to stay. More than a life sentence in the woods. It was here, amidst the horses and the quiet, that Jake had begun to feel the edges of something like peace.

The rain drummed steadily on the roof of the Quonset hut, a rhythmic echo that seemed to match the heaviness in Jake's chest. There was something about the rain that loosened the grip of time, blurring the edges between past and present. Memories surfaced unbidden, fragments of who he used to be. Somewhere in that haze, a line of poetry drifted to the front of his mind. It was faint at first, like a whisper, but it grew clearer as if carried by the rain, Sandburg's words finally surfacing: *Let us write of olden, golden days and hunters of the Holy Grail and men called 'knights' riding horses in the rain.*

Once, Jake had believed in the nobility of sacrifice, the honor of the warrior's journey. He had imagined himself as the knight, fighting for his daughters, Rose and Sarah, his mother, even his ex-wife, Meredith. But now, the image was distorted, muddied by the weight of years and the cost of service. He had fought for honor, for a cause that had cared little for the men on the ground or the lives it shattered. Instead of cleansing him, the rain only soaked him with regret, each drop heavy with guilt and failure.

The thought of his daughters burned brightest. He had wanted to be their hero, their protector against the storms of life. But all he had brought home was brokenness. The question he couldn't silence pressed against his thoughts: *What was it all for?*

Jake closed his eyes, focusing on the sound of the rain, counting his breaths like Addy taught him. *In through your nose... hold... out through your mouth.* The memory of her voice anchored him, calming the tide that threatened to pull him under. She had been the first person to see beyond the layers of guilt and anger, beyond the Marine, beyond the felon. She saw him. And that terrified him.

Jake opened his eyes, his gaze drifting to the southern pasture. The horses moved with deliberate grace, heads dipping to the wet grass, tails flicking like metronomes to the rain's rhythm. Addy had told him once that horses could sense a storm before it arrived, that they absorbed the energy around them and reflected it back. She had said it matter-of-factly, her tone calm, but Jake had seen the way she looked at the horses, the way they responded to her. There was something unspoken between them, a connection deeper than words—a language Jake hadn't yet learned but desperately wanted to.

He thought about the first time he had met her, months ago, in the barn. Sandy had introduced her as his daughter, and Jake had been too worn out to offer more than a curt nod. She didn't flinch. She didn't ask questions or try to pry into his story. She just watched him with the quiet steadiness of someone sizing up a horse—not with judgment, but with understanding. And in the months since, she had somehow managed to slip past his defenses, not by force, but by being there. By teaching him to breathe, to listen, to let go.

"Just like Addy taught you," he murmured under his breath, the rain blurring the edges of the world around him. He wondered how she carried it all—her own pain, the struggles of The Sanctuary, the weight of knowing the brokenness of every person who came here seeking solace. She had told him once, in her usual understated way, that healing wasn't about fixing what was broken—it was about learning to carry the pieces. He hadn't fully understood it then. Maybe he still didn't.

Jake sighed and leaned forward, resting his elbows on his knees. He felt the worn spine of the book beneath his palm, its wisdom something he was only beginning to grasp. *Become totally empty. Let your heart be at peace.* Addy had been the one to read those words aloud to him the first time, during one of their sessions in the barn. She had said it with the kind of calm that made him believe she understood the depths of what it meant. She had grown up here, amidst the mountains, the horses, and the rhythms of life. Her connection to The Sanctuary was palpable, like she was part of the land itself—rooted, steady, unyielding, yet still gentle.

He hadn't told her how he felt, how much she meant to him. It wasn't just admiration. It was something deeper, something terrifying. She was everything he wasn't—whole, strong, resilient. He didn't know how to give her the words without sounding like a coward or an impostor. But he knew one thing: Addy was his anchor. The Sanctuary might have saved his body, but Addy was slowly piecing his soul back together.

Movement caught his attention up by the barn. As his eyes shifted into focus, he saw Sandy's dog, Mindy, trotting through the puddles toward the Quonset hut. Her coat was slick with water, but like the horses, she didn't seem to mind. When she spotted Jake sitting on the small patio, she broke into a run, tongue lolling to the side in what could only be described as a big dog smile. She bounded onto the patio, her front paws landing squarely on his lap, leaving two muddy prints on his jeans. Before Jake could react, she shook herself violently, spraying water and mud everywhere.

"Thanks for that, Mindy," Jake muttered, laughing despite himself. The dog sat at his feet, her head tilted expectantly. "Hold on, girl. Let me see if I've got something for you."

He returned moments later with a peanut butter cracker. "Here you go," he said, handing it over. She crunched it eagerly, then looked up at him again, her tail wagging. "Sorry, that's all I've got," Jake said, stroking her damp fur. Mindy seemed content, resting her head on his lap as he absentmindedly scratched behind her ears. The rhythm of her breathing, the warmth of her presence—it was grounding. But as Jake's hand moved over her coat, the sensation of wet fur triggered something deep in his memory. The world around him began to blur, and he was no longer on the patio...

The rain dripped from the rim of Jake's Kevlar helmet, mixing with the sweat on his face. His M16A4 rifle was at the low ready as he moved cautiously through the rubble-strewn streets. The air was thick with the stench of sulfur and decay, the rain doing little to mask the smell of death. Every step was slow, deliberate, the weight of his gear pressing down on him.

Up ahead, he saw a man kneeling in front of a pile of rubble, his hands clawing at the debris. Jake shouted, "Hands up!" but the man didn't respond. He screamed louder, "Get your hands up!" The man turned, his face streaked with mud, his eyes hollow. "Shoot me," he cried, his voice breaking. "Shoot me!"

Jake's stomach churned as two Marines pulled the man away, his screams echoing in the rain. The interpreter explained that the man's family had been inside the house when it was bombed. His wife, his children—all gone. The man's words haunted Jake: "Kill me like you killed them."

Further down the street, Jake saw bodies—bloated, burned, scattered among the rubble. The collateral damage was staggering. He tried to focus, to push the nausea down, but then he saw them: three stray dogs fighting over a human leg. Without thinking, he fired a shot into the rubble, and the dogs scattered.

One block over, he found another dog with a badly wounded leg. Shrapnel. The animal lay still, its eyes locked on his. Jake knelt, the smell of wet fur filling his nostrils. He looked around for the sergeant. "Where are all these dogs coming from?"

"They're everywhere," the sergeant said, shaking his head. "Nobody takes care of them. You want to help that dog? Put it out of its misery."

Jake stared at the dog, his rifle trembling in his hands. The animal didn't flinch, didn't plead. It simply looked at him, its chest rising and falling in shallow breaths. In that moment, Jake saw himself reflected in the dog's eyes—broken, resigned, waiting for an end.

He pulled the trigger.

Jake blinked, the memory fading as Mindy shifted at his feet. She looked up at him, her eyes full of trust, her tail thumping softly against the patio floor. He let out a shaky breath, his hand still resting on her fur.

The rain dwindled to a faint drizzle; its soft patter now barely audible against the corrugated roof of the Quonset hut. A gust of wind stirred the damp air, sending ripples through the puddles scattered along the path and rustling the leaves of the towering rhododendrons. Jake looked out across the yard, where Sandy's flagpole stood tall, its tattered flag snapping sharply in the breeze. Jake squinted, watching as the wind lifted the Stars and Stripes, its faded fabric catching the faintest hint of light breaking through the clouds. Sandy had raised the flag at 0800 sharp that morning, as he did every day. He always saluted when it went up, his movements crisp, even now, as a man whose hands shook with age. Jake had seen him do it countless times and yet never asked why. Why keep the tradition alive in a place like this, where the echoes of war were both unavoidable and unwelcome?

He stood, the weight of the unspoken question settling in his chest. He didn't know what the flag meant to Sandy anymore—not really. Did it remind him of brotherhood, of duty, of all the men who hadn't come home? Or was it a simpler gesture, an act of reverence for something that had once held meaning, even if that meaning had shifted with time?

And what did it mean to Jake? When he was younger, the flag had been a banner to stand under, a thing that made him feel strong, like part of something greater. Now, it felt more like a weight—a responsibility that had been passed to him but one he never fully understood. It wasn't about the pride of serving. It wasn't about waving a flag or chanting slogans. It was about the quiet sacrifices, the lives lost, the pieces of himself he had left on foreign soil, and the pieces of others he couldn't save. Service, he realized now, was the hardest thing to bear: to love your country enough to fight for it, but to also carry the grief of all that it cost you, and others.

The wind carried the flag higher for a moment, stretching its edges wide before folding it back into itself. Jake frowned, his hands gripping the railing as the dull ache of regret and disillusionment settled deeper. The flag had once been a shield, a source of pride. Now it felt like a hollow echo of something long lost.

The rain stopped leaving behind a damp quiet that settled over The Sanctuary. The air was cool and heavy, carrying the fresh scent of earth and wet grass. Jake took a deep breath, letting the stillness steady him. Mindy stretched at his feet, shaking herself once before lying back down with a sigh. Jake's eyes wandered to the pasture, where the horses stood scattered, their coats slick from the rain. Pepper, the closest to the fence, pawed the ground and let out a low, impatient whinny.

The sound pulled Jake from his thoughts. He knew that call—it was the horses' way of saying they'd had enough of the rain and were ready for the comfort of the barn.

He shook off the lingering haze of his memories. The horses needed him, and he welcomed the task. There was something grounding in their presence, something unspoken and ancient. They didn't ask questions, didn't demand anything beyond the simple rhythms of care—food, water, shelter. In their quiet way, they gave more than they took.

"Mindy, you coming?" he called ducking inside the Quonset hut to return the book to his nightstand. When he emerged with a lead rope in hand, Mindy trotted ahead, her tail wagging. Together, they made their way toward the gate.

Jake opened the latch and stepped into the pasture. The horses lifted their heads, ears swiveling toward him as he approached. Pepper came forward first, her breath warm against his hand as he clipped the lead rope to her halter. The others followed at their own pace, moving in the calm, deliberate way Jake had come to admire. He led them up the trail toward the barn, the tension in his chest beginning to ease. The steady rhythm of hooves on the damp ground, the soft snorts and sighs of the horses—it was enough to pull him fully into the present. The world outside, with its noise and chaos, seemed distant here. The horses' steady movement, their presence grounding him, allowed his mind to slow, to breathe. The weight of his thoughts, the questions he couldn't answer, the worries that had followed him like shadows— they felt lighter in this space.

After the horses were settled into their stalls for the evening, Jake sat down on a bench by the edge of the barn door, Mindy at his feet. He had just finished brushing down the horses, his hands rough from the work, but the familiar rhythm comforting. The storm clouds were parting, revealing a dusky sky that was clearing but still heavy with the last remnants of the storm. He could smell the earth, moist and alive, and hear the quiet breathing of the horses. He reached down and rubbed Mindy's head. She shifted under his touch but didn't move. She was content, and for the first time in a long while, Jake realized that he could be too.

"I'm still here," he whispered. "I'm still breathing." A small, sad smile tugged at the corners of his lips. Not everything needed to be understood or fixed. Sometimes, just breathing through the pain was enough.

The horses shifted in their stalls behind him, heads low in the grain buckets, grazing contentedly. Even the soft whinnies of the horses didn't bother him. They weren't asking anything of him. They were simply living, doing what they did, existing in the moment, in the quiet rhythm of life.

Maybe his service wasn't a thing to be spoken of. Maybe it didn't need to be sung or paraded around for all to see. Maybe it was simply what he carried inside, the weight of it resting heavy and warm in his chest, knowing that the sacrifice was his to bear. He wasn't sure he would ever fully understand it—but maybe that was okay.

Bob Ritchie is a veteran of the United States Coast Guard. After serving, Bob embarked on a twenty-five plus year career in law enforcement, serving for 10 years as an FBI Bomb Technician. In 2005, Bob was assigned to Joint Task Force 76, Combined Explosive Exploitation Cell in Bagram, Afghanistan. Bob currently serves on the faculty at Appalachia State University as a Lecturer of English.

CHAPTER 26

THE BEST OF TIMES, THE WORST OF TIMES

BY THOMAS MORRISSEY

Twenty-two: The number of daily occurring veteran suicides the U.S. Department of Veterans Affairs (VA) studies show. Some call it misinformation. I call it the result of "war with the threat of death removed." Recently, I found myself studying that Greek warrior Ajax. Ajax, who once boasted he had saved himself "without divine assistance." Who, like Jess, the female drone pilot in the opera "Grounded," (If you missed seeing it, you should) also elected a form of suicide as the only path to self-preservation.

Through deliberation, Jess contrasted her struggle—balancing life and war—to that of Odysseus ... "*The Odyssey*" (a word ending in the same last four letters as my last name by the way... lol), asking how different a story *The Odyssey* would have been had Odysseus been able to return home nightly to the comfort of his own bed rather than be stranded at sea, battling the Sirens as it were. In any case, the battle is "not against flesh and blood" but "against principalities, against powers, against the rulers of the darkness, against spiritual hosts of wickedness" as they say. (Ephesians 6:12)

We are in an era in which many or most we know have not even experienced or observed simple "roadkill" with all six senses. Buffered from reality. Physical reality to many is abstract, just a video game as we see in the Jess's "Sensor" who, in the opera, remained nameless I might add.

War, as we who have endured it in person know, entails having skin in the game, risking life and limb. There is no "zona pellucida," zone of protection. War with the "threat of death removed" becomes, as Jess realized, a dysfunctional routine, a numbing, nonsurvivable exercise…22!…. But I digress….

Charles Dickens, in his 1859 novel *A Tale of Two Cities,* wrote: "It was the best of times, it was the worst of times." I guess it depended on just what "times" you actually experienced… when and where. The "best" or the "worst:" a dichotomy worth considering, a point of view at any rate… On to my story….

It was early April1970 when the chartered aircraft's door swung open and a wave of the hottest, muggiest, stench-filled air blasted throughout the cabin. A field-grade type officer appeared in the doorway commanding: "Give me quick-time gentlemen, directly into the buses, don't stop to look around. You'll have 365 days for that." (expletives deleted) We literally raced onto the waiting buses.

Ventilation in these buses was provided through open windows, covered with machine guard type open-wire mesh, "to keep people from tossing hand grenades into the bus," we were told. As we rushed, we passed, lined up in some very loose formations, a rag-tag group of around 100 very "short" guys, mocking us as we passed. Preparing to board their "freedom bird" flights home, to the "World," as it were, getting their last jabs in at the expense of us "FNGs." Welcome to the People's Republic of South Vietnam, enjoy your stay. It was there, I believe, my journey began.

(Image courtesy of the author)

Twelve months later my tour in Vietnam as a helicopter pilot had ended. Returning to the "World" seemed all too strange and unfamiliar. Even the Sears Roebuck in town where I had worked only months earlier seemed a bit abstract in contrast. I mean, riding an escalator, passing another kid I had known in high school, now employed there.

"I heard you were killed" he said as we passed each other. I don't remember which of us was going up or down. Doesn't really matter in the greater scheme of things. "Don't mean nuthin'" as we learned to say regarding just about everything. "It is what it is." Was what it was.

Only days before that encounter, (literally folks) I had been maneuvering my UH1D, Huey Iroquois, Bell Helicopter in and out of a variety of "missions" as they called them. Some boring... tedious at best. Others, real nail-biters I guess you could say. Comic relief if there ever was. Some missions were to where there were, people back home were told, "*no American forces.*"

That kid had no idea. Despite being the same age, I was so much older then than he might ever be... He had a job where the customer was always right... I had earned a living pissing people off enough to shoot at me. Aircraft Commander, 117th Assault Helicopter Company, "Warlords."

I had endured a couple of the strangest years of a life that only those who had experienced it could even imagine. A year in training and a year applying that training. Becoming a journeyman of my trade, (36 combat air-medals), only to be laid off in the U.S. Army's big reduction in forces (RIF) of over 6,000 pilots in 1971. Able to leap tall buildings in that proverbial "single bound." "It's a bird, it's a plane…?" No, neither… It's just me… Tommy Morrissey, back in Miami. On the escalator in Sears. Like it all never happened. A stranger in a strange land.

Others had said: "Haven't seen you around," or ask where I'd been. Like they could understand. Like they really even wanted to know… Me—in my short haircut, amid an ocean of what one songwriter called "overfed long-haired, leaping gnomes." "Hippies." And there I was: totally "out of pocket" by any number of generational definitions. Growing my hair too—just to fit in—if for no other reason.

I seemed to be surrounded by the mundane of everyday life. I mean, Vietnam was very cool in so many ways. A 24-hour a day high. Running on all eight-cylinders. Nine, if we could have. Eyes in the back of our heads. The 1,000-yard stare. Total adrenaline rush, constantly…. Even "boring" was not really *boring*! It was "The Nam." A particular point of view, a perspective gained only through activities like packing up a close buddy's things to send home (Killed in Action or KIA) then going to the O-Club for a few drinks, some strippers, and a band. Bizarre, you say? Nope, normal, just normal day to day.

Day after day. Same ol', same ol' Vietnam. My wife asks, "How was it that nobody prayed?" "Fell to our knees." Today I ponder that very same question. I guess we just didn't think to at the time. There was a safety-wired nut threaded atop of the main rotor hub holding it all together. It was called the "Jesus-Nut." We did check with <u>it</u> daily however.

I had just spent a couple years of my life acclimating myself into a virtual civilization if you will. Not a cult, but a society, a culture. A society based on a strict up/down rudiment in which nothing could really even be questioned. They would say "jump," we would only ask "how high?" as we leaped, praying to not come back down until they told us to.

No waiting around to ponder life or its meaning or even the lack thereof. "Fire for effect." "Drop and give me thirty." "Yes, drill sergeant." "No sir." "Get me a body count." A set of coordinates. A hill number. All reduced to motor-skill actions/reactions. Brief radio transmissions, "Roger that," "fire for effect," keeping it all in the "green arc," watch the RPM, the fuel gauge, oil temp, watch that exhaust temperature gauge. Pay attention to that torque meter. Check for artillery. Friendly fire? What color tracers? Pop smoke, what color smoke? Free fire zone, kill zone…. "Charlie!" Lock and load…. Skills finely honed to precision. Act now, think later. And think later we did.

Michael Herr, in his 1978 book *Dispatches* made mention of a Black marine he had heard about during heavy shelling at Con Thien who'd said, "Don't worry, baby, God'll think of something.".… God always does. "Kill em all, let God sort em out." That's what we said.

Herr also wrote about "Flip religion." He wrote: "It was so far out, you couldn't blame anybody for believing anything…Guys stuck the ace of spades in their helmet bands, they picked relics off of an enemy they'd killed: a little "transfer of power." They carried around five-pound bibles from home, crosses, St. Christopher's, Mezuzahs, locks of hair, girlfriends' underwear, snapshots of their families, their wives, their dogs, their cows, their cars. Pictures of John Kennedy, Lyndon Johnson, Martin Luther King, Huey Newton, the Pope, Che Guevara, The Beatles, Jimi Hendrix…." The list goes on.

"*Wiggier*" then Cargo Cultists, Herr observed. (Those superstitious Pacific Islanders during WWII.) "One man," Herr wrote, "was carrying an oatmeal cookie through his tour, wrapped up in foil and plastic and three pairs of socks. He took a lot of shit about it. His fellow grunts would tell him, 'When you go to sleep, we're gonna eat your

181

f-ing cookie,' but his wife had baked it and mailed it to him, he wasn't kidding." I guess, eating it was not an option. It was holy, an amulet of sorts.

"On operations (Herr continued) you'd see men clustering around the charmed grunt that many outfits created, who'd take himself and whoever stayed close enough through a "field of safety," at least until he rotated home or got blown away, and then the outfit would hand the charm to someone else. If a bullet creased your head, or you'd stepped on a dud mine, or a grenade rolled between your feet and just stayed there, you were magic enough...."

I flew with "that guy," a door gunner who fit that MO. "Jinks," oddly enough was his name. Seriously.

On one mission he began complaining that his foot suddenly developed a burning sensation. We had just completed some insertions and were, at the moment, commuting, safe and sound above 1,500 feet. I told him to check it out. He removed his boot only to discover a Soviet 7.62 round jammed between two toes. Needless to say, Jinks became my good luck charm. I opted to have him as a crew member as often as I could finesse it. Take every opportunity, cross every T.

And "esprit de corps." There's one for the books. Keeping the enthusiasm up. Oh, how we did that. We created silly little nicknacks, cartoonish symbols, artwork...fetishes they were called by those detached from our reality...intellectual observers. We did what we did. I flew Navy SEALS wearing blue jeans into combat. Had it been another war, we would have been cultural icons. Instead, we were called "Baby Killers." We did what we could do. Had to do.

I inherited a helicopter with the portrait of Little Annie Fannie on the nose...the avionics hatch cover...the front of every UH1 "Huey." Our platoon insignia. Our platoon got reprimanded one day by an over-pious bird-colonel who made us paint over her exposed nipples. But still, we in Vietnam had the best humor of any fighting force in history I would bet. What could they do, "send us to Vietnam?"

With our crudely illustrated, unauthorized unit patches: sporting skulls wearing berets, knives in their teeth, accompanied by aircraft sporting Snoopy atop of his doghouse firing rockets into cowering swarms of VC, Yosemite Sam, Porky the Pig, or some other childhood recollection, we were in for the long haul.

Cobra gunship units with names like "Widow Maker," or "Death on Call." There was the 57th "original dustoff" unit whose unit insignia stated, "If you ain't dying, we ain't flying." Or the 159th Dustoff, "The louder you scream, the faster we come." If nothing else, we made the worst of times better (well, maybe not "best") but better. None of that "This We'll Defend" BS!

No, not Woodstock or any R. Crumb style "Keep on Truckin,'" "Mr. Natural," or "Fabulous Furry Freak Brothers" cartoons could hold us back... thwart our creativity among the gore of reality. Janis was singing to us whether she knew it or not. "Take another little piece of my heart." And, that she did.

I have a pretty decent collection of Vietnam War unauthorized aviation unit patches. Admittedly, a few are reproductions. But most are original, from the war. Hand sewn. Flip them over, you can see the hand stitching. "Mama-son" made them. Sold in small unit base camp stores alongside of fake Zippo Lighters, VC Sandals, bootleg Cool Filter King or Winston cigarettes. A little "short-time" in the back room if someone were so inclined. Mama-son sold it all.

I recently saw where Tim Page, the most renowned combat photographer of the Vietnam War, will have his images complete with contact sheets from the 35mm Nikon FTN he carried released in 2025 in a book titled *Nam Contact*. As a photographer, I've already ordered my copy. The publisher, Artbook's, promo states:

Nam Contact harks back to an era when thirty-six frames on a roll of film had to tell the story of a particular action. This book is Page's intricate look at his contact sheets and single images from those sheets... It also contains letters from some of the most noted journalists of the time and further ephemera from what became known as the "first media war" and the first and last war without media censorship." I too shot a hundred or more rolls of 35mm film there, my own personal war.

While documenting the Vietnam Memorial, I met Page a few times there in Washington, DC ....1982, 1983 or 4. Standing atop the wall's apex, we were both observers/participants, I took his photo, it's in my book. We talked briefly. What was there to talk about... really? The Red Sox, the nice day? We shook hands compared cameras and went on our ways. Tim Page died August 2022. He had liver cancer. Another casualty of war.

Tom served as a helicopter pilot with the U.S. Army 117th Assault Helicopter Company (Warlord 23, 1970-71): Aircraft Commander/Mission Controller, flying night "Fire Fly" missions, often into Cambodia and Laos following the Ho Chi Minh Trail. Professor Emeritus, he taught at several colleges and universities throughout his five-decades long career. He was awarded a Kellogg Fellowship, a Fulbright Scholar award to Vietnam and participated in the International Sculpture Symposium, Hue, Vietnam 2002.

CHAPTER 27

A COLLECTION OF POETRY

BY RICHARD ERIC JOHNSON

Dark Corner

medic soldier
martyr saint

never

let them see
you cry

vision blurs
swallow choke
stiffen nerve

tears
later
time and place

here and now
walk away

Descendant of Abraham
The sailor watches
intimidating somber
gray green waves
white capping immensity

sudden sea birds
floating leaves
volcanic peak in sight

nostrils sense
organic earth
fresh water fruits

grains of sand
scattered stars
horizon prayer

the sailor watches

MY COMRADE IN ARMS

(for Rodney and Richard—Johnson)

from a schooling
comrades were Russian commies
from an education
comrades were soldiers

basic training during Nam
we were bunk mates
no dna brotherhood
and the band yet to play

bonds of
practicing weaponry
bonds of
weekend brothel boozing

time of danger
brewing on a far horizon
time of reality
mortality at hand

flesh and blood
torn and flowing

I remember your face
those times from pictures

those times of pride
friendship and toasting
roasting in jokes
laughing arm in arm

decades later
I finger touch
your name
on a granite wall

Four PTSD Recovery Poems:

No Lonesome Valley Here
look up come down
look down come up
playground for gods
devils play as well
angels assemble
choose sides
rituals processions
ceremonies recessions
liturgies hymns
congregate
blessed or cursed
pray and prey
as in the beginning
as in our now
tomorrow still
our will be done

Gather the Flock
stained glass
mosaic colors
gilt crosses

relics
roses
pontificate holy holy

sad day frown
glad day smile
out the door embrace

another day
forgiving
praying

in dreams
heavenly choir sing
allelujah

lay we sheep
all that up and down
to sleep

Collective Gathering
enlightened tribal
Afghan leader
called Americans
the greatest tribe

that was stated
in newsprint
his name
the statement
nowhere on Google
Wikipedia
now to be found

he saw what we seem
interestingly
not to see

drawn from heritages
crossing many waters
cultures ethnicities
we rainbow neon glow
singing dancing
arguing loving

our time as tribe

A Veteran's Senior Moment
I desire
warm cold water
falling rushing
night day
clear vision
storm or calm
I need
books on shelves
paintings on walls
clean satin sheets

feather pillows
windows with a vista
I dream
see your face
try to remember
your name

Richard Eric Johnson lives and writes poetry in Arlington, Virginia. He has authored five full-length poetry collections, and his poetry has appeared in numerous online and print journals. Eric is also a Pushcart nominee. He most recently was honored to be archived at La Salle University's Connelly Library. He is a graduate of Indiana University with a B.A. in Germanic Languages and an M.S. in Education. After a tour in Vietnam and West Berlin, he embarked on a career as a public servant and is now very happily retired.

CHAPTER 28

CONNECTING THE DOTS

BY VALERIE ORMOND

We all have memories of that one assignment, the one where it all seemed to come together. Mine was the Joint Intelligence Task Force – Combating Terrorism, or JITF-CT. The mission, the people, and the drive were unmatched by any other tour during my 25-year Navy career.

Like many of my generation, the Global War on Terrorism was my war. I watched in horror on September 11th when the twin towers collapsed, understanding the world had changed forever. The instant the plane hit the Pentagon, I knew I had lost friends and colleagues, having departed that building less than three months before. Flight 93 crashed 100 miles from where I was stationed. I was a Naval intelligence officer, and intelligence would be key in the arsenal to fight this enemy who had struck hard and by surprise.

Not long after, when I received orders to the Joint Chiefs of Staff (JCS) for Intelligence (J2), I feared I was headed to a dreaded staff job. Fate intervened, and I was vectored to JITF-CT, a new task force aligned to the joint staff.

I didn't know it at the time but found out later that one of my long-time mentors, U.S. Navy Vice Admiral "Jake" Jacoby, had the foresight to see defense intelligence needed to switch tactics after the USS *Cole* bombing. He proposed a major transformation to the Department of Defense's (DoD's) approach to terrorism analysis and operational support and eventually led the stand-up of what became JITF-CT.

After 9/11, DoD and the Joint Staff expanded the previous Defense Intelligence Agency (DIA) counterterrorism office to establish JITF-CT's all-source intelligence fusion center. According to a briefing to then-Secretary of Defense Donald Rumsfeld, the command's mission was to "Generate actionable intelligence to drive planning and operations by exposing and exploiting terrorist vulnerabilities to Deter, Disrupt, and Defeat Terrorists…." In late September 2001, planners envisioned the existing DoD counterterrorism effort to grow from 150 to 700 people—quickly.

DoD terrorism analysis intelligence shifted from "who cares" to "everyone cares" at Mach speed.

In his Congressional testimony regarding intelligence community activities before and after 9/11, Admiral Jacoby summed up much of what would eventually be JITF-CT's mission.

> In our search for relevant information, we must cast a much wider net and then more rigorously mine, manipulate, and interpret the take. In terms of the now-popular analogy of "connecting the dots," we must assume that some of those "dots" are to be found in the observations of gate guards, investigations of thefts and break-ins, or the seemingly benign conversations between terrorist supporters and sympathizers. We simply cannot allow a "dot" to be overlooked, regardless of where it might be found or how deeply imbedded in noise or obscured by faulty assumptions about its nature and relevance.
>
> At its most basic, intelligence analysis is a relatively binary process wherein evidence—observed, reported facts/activities—is combined with assumptions—analytic insight, knowledge—to create an assessment. In essence, the terrorism analyst's job is the extraction of "meaning" from incomplete

evidence, using knowledge, experience, expertise and insight to compensate for absent evidence and ever-present ambiguity.

From the moment I arrived at the task force, the energy, enthusiasm, and camaraderie were palpable. People wanted to be there and wanted to be part of it. I was on fire.

And with fire, sometimes comes rain. I was assigned to the watch, as a Senior Watch Officer on rotating shifts, rather than where I thought I would be going, which was managing the watch. Although disappointed, I followed the old "needs of the Navy" mantra and made the best of it. I took the slower times like midnight shifts to continue to do deep dives into the many aspects of a massive worldwide terrorism mission. Analyzing current intelligence traffic and having the responsibility for real-life terrorist threat indications and warning turned out to be an excellent educational experience.

My watch teams included enlisted, junior officers, civilians, and contractors. Most of the civilians were new to the government, and for many, they were in their first jobs out of college. This can happen when an organization grows from a 150 to 700 people in a few short years. What these young analysts lacked in experience, they made up for in intellect and motivation. I appreciated having the opportunity to lead them and learn from them.

Being in this watch position at the bottom of the totem pole gave me insight into people in the command. In general, I've found that those who treat you poorly when you are at the lowest rung of an organization show you a lot about who they are. And those who treat you with respect while serving in the same position show you who they are, too. It's nice to know who is who right away.

After some time on the watch, the Director for Operations scheduled me to see her. For military folks, when a senior officer wants to see you with no explanation, the mind goes into overdrive. Was I fired? Deployed? Transferred? I'd have to wait and see.

Navy Captain Liz Train was the Director for Operations, and she and I had been colleagues for ten years. Both Naval intelligence officers, we had worked together well in past assignments. But she was my boss's, boss's, boss at this particular time. She had also been in on the ground floor of establishing JITF-CT. She was one of the hardest working and most competent officers I had ever known. Senior leaders recognized her talent, which is why she was in the center of the action.

Liz began, "I know you've been frustrated with your watch assignment here, and we appreciate your patience. We've been working on a restructure of the organization, and it's about to happen. We are standing up the Exploitation Division, and you are going to be its first Division Chief." I didn't know what an Exploitation Division was, but I would soon find out.

I said goodbye to my watch team and continued to see them and was happy to follow their careers as they rotated off the watch floor to daytime analyst positions. We would always have that camaraderie of having been a team, working odd hours, weekends, and holidays, and figuring out intelligence problems when no one else was around. Twenty years later, I am still in contact with Watch Team Four colleagues.

One of the factors that separated JITF-CT from other organizations was its cast of characters. No one had to be told to think out of the box, in fact, it was more often people had to be reminded to color within the lines. Everyone wanted to fight hard, but we were also working in a field with ever changing rules. We had to think ahead of the enemy because reacting was too late. We knew our mistakes could be costly...to those in the field and their families...and to those in the homeland we failed to protect.

Starting at the top with our Director, Jeff Rapp, a retired U.S. Army intelligence colonel, the command embraced innovation, a "why not" attitude, and an understanding that we would have to test new concepts to succeed. Not all ideas would work, but we could not afford to be 100% risk averse.

One of our star counterterrorism analysts in her first job was Christy Abazaid. Christy went on to eventually become the senior executive Director of the National Counterterrorism Center (NCTC). In an interview with the Office of the Director of Naval Intelligence, she shared her experience about her start at JITF-CT.

"It was almost like a start-up mentality…and it was just a lot of people being scrappy about how we're going to do this whole fight against al-Qaida, and fight against a range of terrorist threats we were dealing with…."

We had some old school salts from the previous terrorism outfit that JITF-CT had become. One was famous for a series of his famous sayings, one of which was, "I always say if you're not pissing someone off, you're not doing your job." This reflected the sense that the command overall understood each one of us could not continue with the status quo and expect to make the needed changes.

When I stood up the Exploitation Division it grew from what had been a small branch to a full division. Fortunately, the talented former Branch Chief became my Deputy Division Chief. He was the subject matter expert, an Army veteran, and a brilliant analyst and manager. The mission was personal to him. His brother, a New York City firefighter, gave his life in The World Trade Center on September 11th trying to save others.

Exploitation Division, or X-Div, had an important mission. X-Div had three primary functions: detainee and interrogation operations, document exploitation (DOCEX), and watchlisting. Operationalized X-Div processes became an innovative model for information sharing and collector/analyst collaboration. Our three branches worked closely together because information uncovered by one could be useful to another. In a notional example, documents found in a raid in one location in Iraq may have implications about a detainee in holding somewhere else who was claiming to be an innocent farmer. We were connecting the dots Admiral Jacoby spoke about, and many times those dots were spread continents apart from people who seemed to have no connection.

JITF-CT staff lived for the operational support, but they also responded and excelled at the strategic level. They met time critical requirements from the Joint Staff, the Secretary of Defense, Congress, and the National Security Council. We were repeatedly called upon to brief high-level executives on our programs, our processes, our progress, or to address current real or perceived terrorist threats. This team could do it all.

I had the pleasure of leading a team who wanted to deploy and who were talented, smart, and creative. The military were expected to deploy, but our civilians stood in line to endure separations from families due to their loyalty to the mission. The X-Div mission grew swiftly to include teams in Cuba, Iraq, and Afghanistan.

For DoD detainee operations, our job was to improve intelligence support. Our analysts provided interrogators and briefers background and intelligence analysis on those being held. They also fed interrogators and debriefers information on intelligence gaps and requirements that detainees might be able to fill. They reviewed interrogation reports and provided post-interrogation assessments and intelligence reports for the entire intelligence community.

We considered and encouraged every idea. At one division meeting, frustrated by how many "known" terrorists had been captured and released in the field due to troops not knowing who they were, we had an idea. Why not produce terrorist recognition cards (similar to baseball cards) for the people in the field? We ran the idea up the chain of command, had our reserve unit chomping at the bit to support it, and launched the program. It was simple, and we received positive feedback from the field. But these were the kinds of crazy ideas the JIFT-CT crew were not afraid to suggest.

Cover Card for Terrorist Recognition Aid Deck (Courtesy of the author)

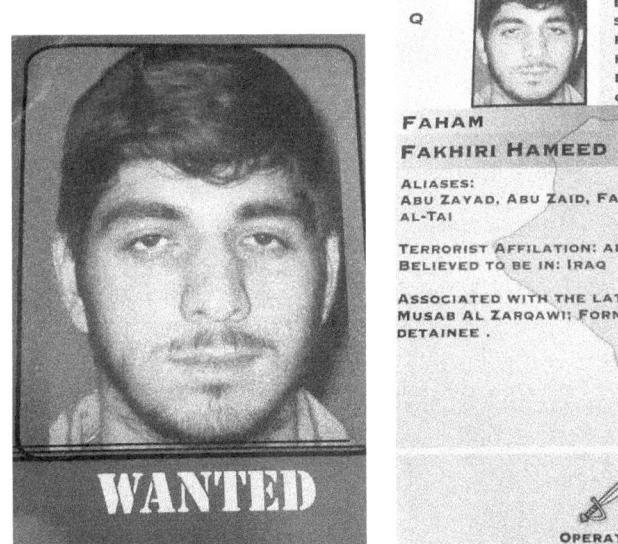

Front and back sides of terrorist operative in Iraq recognition card (Courtesy of the author)

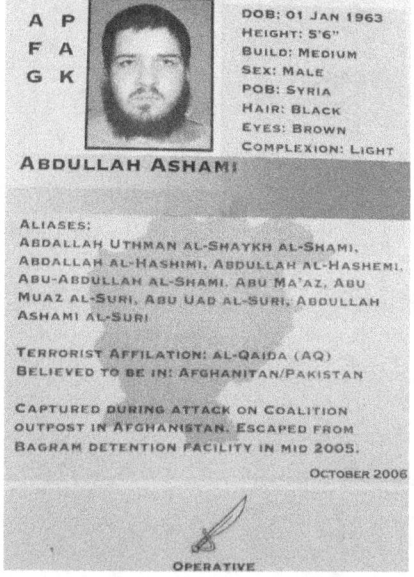

Recognition card to help identify terrorists and escaped detainees in the field—Afghanistan Pakistan region (Courtesy of the author)

We continued to learn the value of integrating document exploitation into detainee operations support. Our DOCEX analysts performed exploitation at the National Media Exploitation Center in Tysons Corner, Virginia, to produce intelligence products. They worked side-by-side with interpreters, many of whom were local taxi drivers because the government didn't have enough linguists in certain languages. Our analysts sifted through the droves of information recovered by our forces overseas to find those nuggets of intelligence that either filled gaps or opened new lines of investigation. Realizing the value of DOCEX, we expanded JITF-CT operations in ten deployed and interagency sites. Our team increased intelligence reporting from captured documents and media and provided real-time support to special operations missions worldwide.

In 2003, DIA tasked JITF-CT with the DoD watchlisting mission responsible for nominating individuals found in DoD information to the NCTC to include in the Terrorist Screening Database. I've mentioned the cast of characters, and here is when another one came into play. My first duty as X-Div Division Chief was to secure resources to support this mission. JITF-CT's Deputy Director had a long history supporting the counterterrorism mission, but his true superpower was on the resources side. He noticed my wide-eyed expression when he started telling me what would be required to obtain the resources we needed, so he patiently led me through the process.

I developed a brief that outlined the process of hiring contractors managed by military and civilians to implement homeland security watchlisting requirements in the time period specified. The charts were staggering, considering the amount of information being collected versus our analytical ability to review it. As a Navy Commander, I briefed this program to the JCS J2 two-star general seeking approval for what started as a 30-person program. He approved it, and the program has now grown to over 200 people today. I would like to think that we kept a few bad guys out.

The first contracting team we hired ended up developing a critical new capability ahead of milestones and doubled DoD's contributions to the national watchlist in three months. Again, in our division's brainstorming sessions, we thought we should include all former and current DoD detainees on the list. It was a watchlist, not a blacklist, and we understood there were some detainees that should have been kept, but who we didn't have the intelligence on at the time to justify it. And some detainees may have changed their opinions of the U.S. after we detained them. We nominated over 50,000 personnel to the national watchlist in the first few years of our mission, receiving intelligence community recognition as a key contributor.

During these successes, Mr. Rapp called me in for an unscheduled meeting in which he said, "I'd like you to be the next Director for Operations."

I said, "Thank you, sir, for your vote of confidence, but we've hit our stride in X-Div and are doing important work. I'd like to stay where I am."

"I'm not really asking you; I'm letting you know what your next job is." Liz was in the office and stayed silent but smiled. She had worked her succession management.

I moved up to the Director for Operations position and found a huge scope of work that I had not even been aware of. I had also been promoted to Captain at this point and found myself third in charge of this incredible organization. Ordinarily, the third person in charge would not be a big deal. But with the requirements to have a senior in so many in person meetings in the Washington, DC area, I often found myself in interesting situations.

On one occasion at the Old Executive Office Building of the White House, I had to represent DoD in an interagency meeting on watchlisting, which fortunately I understood. A senior CIA officer with whom I and my leadership had disagreed with on this subject in the past saw this as his opportunity to talk over the kid in the room while I tried to present DoD's position. I was by far the junior person, and the only person in uniform in the meeting.

Fortunately, the White House senior in charge spoke up. He happened to be a retired Coast Guard officer. He shut down the CIA senior and said, "Let the Captain speak." I will never forget the respect I had for that man and our sea services connection at that time.

I presented the DoD position and eventually won the argument. It may seem trivial, but the issue amounted to massive resource commitments that made it worth a fight. And it was worth taking on that smug senior who later rose to the Cabinet level and has had a rather storied career.

What was it that made this tour and this organization so different? We had capabilities others didn't because of our people. We had the focus, and a team so connected that we could do more than others. We were the envy of other sections of DIA, not only because of our mission and the attention we received, but because of what we had built from the bottom up in reaction to the attack.

I considered Dolores "D." Heib one of JITF-CT's main characters who spent a total of 11 years with JITF-CT, five as an active duty Army lieutenant colonel and six as a civilian in a variety of roles, including replacing me as the second chief of X-Div. She spent tours in Iraq and Afghanistan and was a no-nonsense leader with a sense of humor. She shared her reflections.

The "JITF Way" was a total team effort each and every day where egos did not exist. There was simply something very unique about these counterterrorism professionals. They were humble, introspective, and unrelenting in their mission to both warn American forces of terrorism threats and to provide world class intelligence enabling warfighters to identify, locate, and defeat terrorist organizations. In truth, we were like this even before 9/11. But after?

We had countless long days seven days a week for months on end, and nobody ever complained. It was a total subjugation of everything in our lives outside of our work, and we did it with honor.

We took our work seriously, but we never took ourselves seriously. You simply had to keep your ego in check and take others' perspectives and insights into account because our adversaries were ruthless and cunning. Months after 9/11 when we finally paused to codify our core values to ensure we instilled them in our growing workforce, it came as no surprise we quickly established them as: Teamwork, Commitment to Excellence, Creativity, and The Will to Win. Every day we passionately and effortlessly instilled these values in everything we did. It was an honor of a lifetime to serve with these professionals.

The people of JITF-CT demonstrated a heartfelt commitment supporting a difficult and complicated mission. I'm grateful to have been part of this special startup where lives depended on us successfully connecting the dots.

Valerie Ormond retired as a Navy Captain after a 25-year career as a U.S. intelligence officer and founded her own business, Veteran Writing Services, LLC. Valerie's three novels—*Believing In Horses, Believing In Horses, Too*, and *Believing In Horses Out West*—have won fifteen national and international awards. Her fiction and non-fiction stories, articles, and poems have been published in books and magazines. She proudly serves as MWSA's Vice President.

CHAPTER 29

IWO JIMA VETERAN'S DIARY FOUND: AFTER EIGHTY YEARS, A NAVY SIGNALMAN'S FIRST-PERSON ACCOUNT OF THE BATTLE

BY DIRK PLANTINGA

Eighty years ago, on 19 February 1945, twenty-five year old Signalman First Class Clarence J. Gerwien described his "front row seat" from the bridge of the troopship USS *Bladen* (APA-63) when United States Navy ships began shelling the island of Iwo Jima, before dawn, on D-Day, 19 February 1945.[1] Gerwien watched the fiery blasts and felt the concussive power as USS *Tennessee* (BB-34) fired her guns, sending fourteen inch/fifty caliber shells aimed to soften enemy positions ahead of the amphibious assault.[2] USS *Tennessee* (BB-34) was

---

[1] 1937-1945 - The Times that Tried Men's Souls by Clarence Gerwien (year)

[2] Naval History and Heritage Command. Pre-landing Bombardment and Landing Preparations, history.navy.mil

one of the battleships that survived the attack on Pearl Harbor, as was USS *Nevada* (BB-36). Finally, these mighty ships had the opportunity to attack land belonging to the Empire of Japan.[3] "They (the Japanese on Iwo Jima) sure are getting an awful pounding. The shelling and dive bombing were terrific," Gerwien recorded in his diary.[4]

At daybreak, Navy and Marine aviators of Task Force 58 launched from more than twenty carriers. Planes buzzed just overhead the Navy ships as they closed in for low-level strafing attacks ahead of the invasion.[5] At H-Hour, thousands of soldiers of the United States Marine Corps, along with Navy corpsmen, charged onto the beaches.

The infantry units fought their way across the narrow neck of the island to cut off Mt. Suribachi, the highest ground. But for months prior to the invasion, the Japanese had been engineering elaborate underground defensive positions that included sniper nests hidden in caves, pillboxes, and various underground positions which the bombardment could not completely destroy.[6] With the invasion forces bottled up on the beach, the Japanese unleashed a withering barrage of bullets and explosives from positions embedded on the face of Mt. Suribachi.[7]

Squadrons of fighters and dive bombers, piloted by Navy and Marine aviators of Task Force 58 provided direct air support of the invasion, while the fast carriers focused on air patrol and interdiction of Japanese kamikaze attacks.[8]

---

[3] Naval History and Heritage Command. Pre-landing Bombardment and Landing Preparations, history.navy.mil

[4] 1937-1945 - The Times that Tried Men's Souls by Clarence Gerwien

[5] Iwo Jima: A Tale of Two Carriers, Commander J. Brandon Wilgus, U.S. Navy (Retired), U.S. Naval Institute Naval History, February 2025

[6] US Naval Aviation in the Pacific, The Office of the Chief of Naval Operations, US Navy, 1947

[7] ibid

[8] US Naval Aviation in the Pacific, The Office of the Chief of Naval Operations, US Navy, 1947

On the evening of 21 February, Navy aircraft carriers came under attack by swarms of Japanese aircraft, piloted by Kamikazes. Six enemy aircraft scored five hits on the Navy's fast carrier, USS *Saratoga* (CV-3). Ninety-three sailors were buried at sea, with dozens more MIA.[9]

After the sun set, the escort carrier USS *Bismarck Sea* (CVE-95) was struck repeatedly by Japanese kamikazes. The ship was lost along with 318 sailors[10].

That night, still aboard the USS *Bladen* (APA-63), Gerwien wrote in his journal, "The (Japanese) Air Force attacked us this evening and damaged a carrier nearby. Casualties are coming back in increasing numbers. We have taken several men aboard this ship for treatment."[11]

On 23 February, a patrol of Marines accompanied by Sergeant Louis R. Lowery, a Marine photographer for *Leatherneck* magazine, reached the summit of Mt. Suribachi. The stage was set for a moment in time that was noticed around the world: Americans raising The Stars and Stripes over Japanese soil.[12]

A civilian photographer from the Associated Press, combat journalist Joe Rosenthal, took the iconic photograph of a group of Marines hoisting an American flag into the wind. However, this had not been the first photograph of the first American flag to fly atop Mt. Suribachi. Lowery photographed the first flag being raised, before the civilian press arrived on the scene.[13] While documenting the scene, the patrol came under attack, and Lowery was injured, the camera destroyed, but the film survived.[14]

---

[9] ibid

[10] The Fleet at Flood Tide - James D. Hornfischer, 2016

[11] 1937-1945 - The Times that Tried Men's Souls by Clarence Gerwien

[12] Flags of Our Fathers, James Bradley with Ron Powers, Bantam Books, 2000

[13] Flags of Our Fathers, James Bradley with Ron Powers, Bantam Books, 2000

[14] Morison, Samuel Eliot. *Victory in the Pacific, 1945 – History of the United States Naval Operations in World War II*, Vol 14. (Boston: Little, Brown, 1960): 389.

The first flag, which Lowery photographed, was energetically hailed with raucous cheers from the beach below. A chorus of horns, bells and whistles from hundreds of ships offshore swept over the island, punctuating a symbolic moment.[15]

Gerwien watched the wind whip the flag into full view, but the sailor was under no illusion that the battle was won. Later that evening, Gerwien wrote:

"...had an air raid about 8 o'clock this eve. They came pretty close to us this time." The next day, Gerwien's unit, B4D15, donned their packs, descended ropes attached to the hull of the USS *Bladen* (APA-63), and into a landing craft. "We were equipped with rifles, helmets, full packs, etc., the same as the Marines. Except I was a sailor, not a Marine," Gerwien clarified in his diary entry. At 1500, one day after the flag raising, Clarence Gerwien's unit of Navy signalmen reached Rock Beach and dug in for the night.[16] While the ferocious battle raged on, Gerwien spent a sleepless night huddled in a bomb crater, along with a buddy. The two were dissatisfied with the small amount of cover the shell crater provided and spent the next day improving their shelter. "Furious fighting going on up at the front," he wrote.[17]

After spending three nights in that crater on Rock Beach, Gerwien moved to Green Beach and again spent the day trying to dig into the volcanic crust for shelter. Exhausted, Gerwien stood watch until midnight.

Gerwien wrote a letter to his "one and only," Mrs. Kay Gerwien, who he described as his "precious little wife," or "my honey." There were two things Clarence Gerwien did nearly every day: Read three

---

[15] Uncommon Valor, Common Virtue - Hal Buell and Lou Reda, The Berkley Publishing Group

[16] 1937-1945 - The Times that Tried Men's Souls by Clarence Gerwien

[17] ibid

chapters of the Bible and write a letter to Kay. Classically, letters didn't flow smoothly between them, sometimes arriving in batches, out of order, or none at all for days. Nevertheless, he wrote Kay a love letter each day, though at times expressing frustration that she didn't write to him as much. Not long after, Gerwien would send a letter that included an apology after receiving several letters from Kay that had arrived in the same mailbag.

The duties of a Navy signalman sent Gerwien traversing the island to man the radio shacks that held the equipment needed for command and coordination on and off shore. There was White Beach, Black Beach, Blue Beach, Rock Beach, Yellow Beach, Brown Beach and Purple Beach, among others. One day, Gerwien came across a cave, and inside he found a bicycle, formerly owned by a Japanese soldier. The bicycle made Gerwien's treks around the island a little easier, but along the way he beheld the temporary graves of thousands of soldiers. "It was quite a site to see row upon row of white crosses, all lined up."

Gerwien spent ninety-seven days on Iwo Jima, as the battle went on for months after the flag picture caused headlines.[18] Gerwien and his unit were repeatedly attacked. A sniper killed a man just a few steps away from Gerwien. Explosions from mortar attacks demolished the unit's tent camp. Finally, Gerwien's battle on Iwo Jima finally ended on 1 June. "Goodbye to Iwo Jima" he jotted in his diary. Just before departing the island, Gerwien convinced another sailor to give him $75 for that Japanese bicycle.[19]

Clarence Gerwien reunited with his "one and only" Kate and daughter Cookie. Their family grew with the arrival of their son Harold, followed by sons Alan and Gary. But Clarence Gerwien remained in the Navy Reserves and was called up during the Korean War. Gerwien again reunited with his family, now a veteran of two wars. Clarence Gerwien was one of the heroes that survived.

---

[18] Morison, Samuel Eliot. *Victory in the Pacific, 1945 – History of the United States Naval Operations in World War II*, Vol 14. (Boston: Little, Brown, 1960)
[19] 1937-1945 - The Times that Tried Men's Souls by Clarence Gerwien

During the Vietnam War, Clarence Gerwien's eldest son Harold (Harry) J. Gerwien voluntarily enlisted in the Navy, after receiving his draft notice in 1970. Photographer Mate Gerwien went on to chronicle United States Naval Operations in the sky, on board ships, and from submarines below the surface of the sea. The younger Gerwien and his camera gear commemorated thousands of historic events during the Cold War era. He captured candid images of United States Presidents and traveled the world as staff photographer for three Secretaries of the Navy. Harry Gerwien was named Navy Photojournalist of the Year in 1987, and pictures he took twice made the cover of the *New York Times* magazine. In 1988, Clarence Gerwien's son Harry stepped foot on Iwo Jima himself, accompanying Secretary of the Navy H. Lawrence Garrett. Harry Gerwien only spent a short time on the island where his father survived 97 days. Gerwien looked through the viewfinder of his camera and scanned the area from the same spot his predecessor photographed the flag in 1945. The lens Gerwien looked through struggled to approximate what Lowery or his father saw. "It was hard to envision the enormity of the loss of life," Harry Gerwien remembered.[20]

Having survived ninety-seven days on Iwo Jima, the sailors from Clarence Gerwien's unit made a pact. The last man alive from B4D15 would pop the cork on a bottle of champagne in a salute to the others. Years passed, and then Gerwien and his Commanding Officer became the final two. Gerwien had described his Commanding Officer Roy Bouque as the type of officer that commanded respect because of his "bearing and good treatment of those under him." In 2009, Gerwien received a telephone call from Bouque's daughter, who informed him of her father's passing. In a final expression of her father's goodwill toward his men, she sent Clarence Gerwien a bottle of champagne. He popped the cork, but the teetotaler never took a drink. A veteran of the battle of World War II, the Korean War, and a member of the Greatest Generation, Clarence Gerwien lived 101 years.

---

[20] Interview with Harold J. Gerwien by Dirk Plantinga, 2024

Signalman Clarence Gerwien in Hawaii in Jan 1945 photograph taken a month before the battle of Iwo Jima (Courtesy of the author)

Iwo Jima and Mt Suribachi August 1989 (Courtesy of the Harold J. Gerwien collection)

Harold J. Gerwien on top of Mt. Suribachi in August 1989 (Courtesy of the author)

TUESDAY MAY 8

This is the day millions of people all over the world have been waiting for. Germany surrendered unconditionally ! I received the news this morning from the Signalman on the LST 800. Now to get the war on this side over and then go home for good. A year ago today I arrived in Pearl Harbor with our unit.

Clarence Gerwien Diary entry May 8, 1945 (Courtesy of the author)

Dirk Plantinga is a journalist, researcher, and archivist focused on United States Naval history. Dirk's interviews with Navy veterans preserve their oral histories for their families, future generations, and historical significance.

Dirk is a member of the Military Writers Society of America, Intruder Association, Tailhook Association, Association of Naval Aviation, U.S. Naval Institute, and the VA-65 Reunion Group.

CHAPTER 30

I SAW A "SISTAH" (Sister)

BY BJARDEN HOLTER

I saw a "sistah" today, and
of all places, in Eastern Europe.
She wore a green battle dress uniform, as did I.
The difference was in the flags that we each
wore on our shoulders.
Her allegiance was to one and mine to another.
As we strolled in opposite directions,
her big beautiful brown curious eyes looked back
to sneak a peek at me; as I, too,
peeked at her.
Our eyes locked as we nodded our mutual respect,
But in those quiet seconds that passed between us,
our long ancient history screamed for our attention.
I could feel her pulse pulsating through centuries past
– as she could feel mine.
The soundless drumbeats on an ancient
African battlefield beat louder and louder
as it throbbed in our ears and
threatened to tear our hearts out;

The spasm of excitement convulsing through our blood
rushed to confront us.
Our feet danced impatiently
…itching to move forward toward battles,
so long ago fought.
Then…
the violence of misunderstanding
erupted and manifested in the
spilled blood of our fathers,
our uncles,
our brothers,
our husbands,
our sons …
some justified, most not.
My sistah and I moved in a slow dance,
…remembering the tears that flowed down the
beautiful black faces of our mothers,
our aunties,
our sisters,
our daughters, … all for naught.
For, we remembered our mutual covenant of "never again."
But here we stand, in the army of another
…so far far away from our ancient roots.
She …completely loyal to her new lover and, I…to mine.
My lover is the American flag,
the red, white, and blue; and
hers was the flag of the French,
the blue, white, and red.

In that moment,
our gaze became an uncertainty as to how
our mutual history became lost
…and we became separated.
Separated by a vast ocean of ideology and time,
estranged by wars and
so much spilled blood
that it could fill the oceans a million times over.
We stared confused
…as to how the irony of life had brought us
together in this faraway place
…in a "peacekeeping mission."
Finally, our eyes dropped their gaze.
Then she and I,
both Soldiers, "carried on"
as if nothing happened … but it did.

Bjarden Holter is married to Ron, the mother of two adult sons, a wonderful daughter-in-law and has two beautiful grandsons. She is a retired Army JAG officer, an attorney mediator and works as a part-time magistrate. She enjoys Bible Studies, reading, writing poetry, likes traveling, spends time volunteering in the community and has published a children's Christian book.

CHAPTER 31

SEVENTY-TWO
HOUR ESCAPE

BY WALTER "BUTCH" MAKI

I n March 1968, after my tour in Vietnam and a stint convalescing at
William Beaumont Army Hospital, I was reassigned to Fort Carson, Colorado.

By January 1969, three friends and I had had enough of the freezing Colorado winter. So, we decided to take a three-day pass to my
parents in Phoenix in search of sun and warmth. We calculated it was
about 800 miles, way outside the 250 miles authorized for a weekend
pass.

Marty Zawaky said, "That's a 12-hour drive."

But with my brand-new Chevy Chevelle Super Sport, I told the
guys, "We can do it in 11 hours, tops." After all, what could go wrong
when you've got horsepower, youth, and just dumb enough to make it
interesting?

We were so eager to escape the cold that we convinced the Charge
of Quarters to sign us out at 12:01 on Friday morning, allowing us to
pack up and hit the road seven hours early, at five in the afternoon on
Thursday. Who needs sleep anyway?

The drive started great. The Chevelle was purring, the highway was clear, and we felt good about our escape plan. That is, until the middle of the night, we found ourselves somewhere between Holbrook and Heber, Arizona, in the pouring rain. Feeling invincible, or maybe just way stupid, I was cruising at over 100 mph, feeling my butt leaving the seat at the top of every rise. Then we crested one, and there it was, a flooded wash at the bottom of the dip. There was no time to stop, and slamming on the brakes would've sent us skidding out of control. Bob Bodo let out a scream from the passenger seat that could rival a horror moviegoer while the guys in the backseat woke up in sheer panic.

As we hurtled toward the rushing water in that split second, I thought *I had made it through Vietnam but was killed by a mud puddle that would be in my obituaries.* I braced for the impact. To my astonishment, we somehow glided across the flooded wash like Jesus walking on water.

Once safely on the other side, I muttered, "That had to be divine intervention."

Pale and still catching his breath, Bodo snapped back, "We hydroplaned, you idiot!"

I grinned and replied, "I may be an idiot, but that brave hero of Vietnam sitting next to me with a chest full of medals was screaming like a banshee as if it were coming straight for us."

Bodo sheepishly said, "I thought he was."

Everyone started laughing, causing Bodo to say, "Screw you guys. That Irish female spirit whose <u>wailing</u> warns of an <u>impending</u> death thing almost came to pass," while taking a long drink of beer.

That flooded dip made me realize that whether I was crewing a Huey in Vietnam or driving a Chevelle Super Sport at home, the line between genuine courage and outright recklessness was razor thin.

We arrived in Phoenix at five a.m., and Marty said, "I told you 12 hours."

I was tired and barked back at him, "We would have been here earlier if you and John Taylor didn't drink beer all night, so we had to stop every hour or so for you guys to pee. Then Bodo just had to eat tacos in Santa Fe at his favorite restaurant. A place that, when we went in, had no one there except the owners. I'm sure they were praying we came to buy the place."

Mother and Dad were already awake, waiting for us when we arrived. Mother's face lit up the moment we stepped inside, her eyes brimming with relief, joy, and motherly concern. She reached out, giving each of us a warm smile, and said, "It's so good to have you boys here."

After the introductions and hearing of our water incident, her eyes darted from one of us to the next, lingering just a moment longer as if making sure we were all in one piece. Then, as only a mother could, she clapped her hands together and said with a determined tone, "Go on out to the back porch, get comfortable, and I'll whip you boys up a proper breakfast."

The back porch was a warm, inviting sanctuary, a simple structure with a roof over a concrete pad separating the house from an emerald green backyard offering a shaded retreat from the morning's Phoenix sun. Hanging flower baskets, brimming with bright petunias and marigolds, added splashes of color against the backdrop of a stucco wall. The scent of blooming citrus trees drifted in from the yard, mingling with the rich aroma of breakfast cooking.

Mother had set the table with her best china. A checkered red-and-white tablecloth covered the table, its edges fluttering slightly in the warm desert breeze.

Bodo said, "What is the penalty for AWOL? The weather is great compared to the cold and snow of Colorado Springs. I may not leave on Sunday.'"

"Absent without leave will get you a loss of flight status and a stripe. It also will happen if the CO finds out we're in Phoenix on a weekend pass," I answered.

Breakfast was a feast fit for soldiers. Mom brought out platters of sizzling bacon, crisping to perfection, a mountain of golden buttermilk pancakes glistening with melted butter, and a pitcher of maple syrup. She served scrambled eggs, cooked just right, roasted potatoes, and mingled them with onions and bell peppers for extra flavor.

A pot of strong coffee sat ready, the smell alone enough to perk everyone up for those who prefer some juice. A pitcher of freshly squeezed orange juice, sweet and tangy, rests on the table, condensation beading on the sides.

"Eat up, boys," she said, her voice soft but steady, a mix of pride and love. "You need your strength after that all-night drive."

Dad leaned back in his chair, his coffee mug hovering near his lips as he surveyed the group with a curious smirk. "Did all of you go to Vietnam?" he asked, his tone casual but laced with genuine interest.

I nodded, already sensing where this might go. "Yep. You know I was in Pleiku, Bodo was up in the north, and Marty hung out in Ankhy. We were tangling with the North Vietnamese Regular Army, but Taylor was down in the Delta where they fought Boy Scouts."

"Boy scouts?" Dad's eyebrows shot up as he set his mug down.

"Yeah," I said with a straight face. "You might've heard of them by their other name, the Viet Cong. Farmers by day, militia by night. They probably were awarded merit badges instead of medals."

John, who had been silently buttering a biscuit, looked up, clearly unimpressed. "I guess you think they weren't shooting bullets at us, huh? Just spitballs? You wanna see pictures of the holes in my bird?"

Without missing a beat, I smirked. "Oh, sure. Holes from what? Indigenous arrows?"

The table erupted in laughter, with Marty nearly choking on his coffee. John shook his head, trying to suppress a grin. "You're full of it, you know that?"

"Of course I am," I shot back. "Never denied it. I wish I could have seen those arrows. One of their Eagle Scouts probably carved them."

Even Dad cracked up, shaking his head as he leaned back and waved us off. "You boys sound exactly like my buddies and me in World War II.

After breakfast, the feast had left us pleasantly stuffed. The chairs and loungers scattered across the porch and backyard seemed to call our names, promising a blissful escape into the arms of relaxation.

The Phoenix sun was already warming the air, its gentle rays a welcome contrast to the icy winds and snowdrifts of Fort Carson.

One by one, we claimed our spots. I flopped onto a lounger with an exaggerated groan, patting my stomach. "If I move too fast, I might split my pants," I mumbled, prompting a chuckle from Marty, who stretched out on a cushioned chair nearby.

Bodo found another spot and pulled his hat low over his eyes, mumbling something about "Practicing desert camouflage," while John stretched out on the grass, folding his hands behind his head and grinning at the sight of a sky with no snowflakes.

"This is the life," I muttered, letting my head fall back as a soft breeze rustled orange and grapefruit trees. Compared to the bitter cold and grueling routine at Carson, I thought *this felt like paradise.*

My father shook me awake as the sun dipped low in the west, casting a warm, golden glow across the porch. His voice had that tone he used when something serious was up. "Boys, there's someone here asking for you," he said.

Groggy and still half-asleep, we rubbed the sleep from our eyes and staggered into the living room. What we saw stopped us dead in our tracks. Any trace of fatigue evaporated in an instant.

Standing beside my dad was a Sergeant Major, E9, sporting a crisp uniform, an air of authority, and an MP armband that seemed to gleam in the fading sunlight. His presence was formidable, to say the least.

My dad's face was solemn, his lips pressed tight, as the Sergeant Major spoke in a voice that left no room for negotiation.

"Where's your passes?" he barked, his tone more an order than an inquiry.

I fumbled to hand him mine, my heart pounding as my hands turned clammy with worry. He scrutinized it intently before fixing his gaze on us, his brow drawing into a tight frown. "Do you boys know the limits on a three-day pass?" he demanded.

I managed a shaky, "Yes, Sergeant Major."

"Then why are you boys driving more than three times the allowed mileage? You got an explanation for this?"

Bodo, always quick with a quip, shrugged and said, "We just wanted to thaw out from the bad winter in Colorado."

The Sergeant Major, with arms crossed like an iron gate, smirked. "Maybe you can finish thawing out in my jail," he said dryly.

Marty, the only one with a brain, asked, "Can't you cut some slack for four decorated Vietnam vets?"

Before the Sergeant Major could reply, my mother appeared in the doorway, hands on her hips, her expression of exasperation. "That's enough," she said firmly, pointing a finger at both men. "These boys don't deserve to be scared half to death!"

The Sergeant Major's stern facade cracked, and he chuckled. My dad, unable to keep up the act, clapped the man on his back and grinned. "All right, all right," he said. "Boys, meet our neighbor Harry Hensel. He's in the National Guard, and we thought we'd play a little trick on you."

I exhaled a breath I didn't realize I'd been holding. "Very funny, Dad. I almost had Vietnam flashbacks."

The room burst into laughter, the tension finally broken. After the initial shock wore off, we all settled in for beers and swapped stories. By the end of the afternoon, the whole ordeal was just another tale to laugh about and tell the guys back at our company.

We basked in the Arizona sun all day Friday and Saturday, savoring every golden ray like a rare treasure. Stretched out lazily, soaking up the warmth that felt like a blessing after enduring the icy grip of Fort Carson. At night, we hit the clubs, swapping the quiet hum of the desert for pulsing music, laughter, and a few too many rounds of drinks.

Sunday came all too soon, and with heavy hearts and maybe a touch of a hangover, we bid farewell to Phoenix. We hit the road before noon, determined to return to our duty station without pushing our luck or the speedometer past ninety. The open highway stretched before us, and though the snow and cold awaited us at the other end, the memories of the sun, laughter, and a touch of panic stayed with us for the journey.

Butch Maki, a retired veteran, lives in Los Lunas, New Mexico, with his wife of 30 years, Patty. Since retirement, Butch has embraced writing as a creative outlet. His debut novel, *Bikini Beach*, has earned multiple awards, and became an Amazon bestseller.

www.ingramcontent.com/pod-product-compliance
Lightning Source LLC
Chambersburg PA
CBHW051106030726
47504CB00006B/1809